Robert J. Scully
With Endnotes by Marlan O. Scully

The Demon and the Quantum

Robert J. Scully
With Endnotes by Marlan O. Scully

The Demon and the Quantum

From the Pythagorean Mystics to Maxwell's Demon and Quantum Mystery

Second edition

WILEY-VCH Verlag GmbH & Co. KGaA

The Authors

Robert J. Scully
1707 English
Irving TX 75061
USA

Marlan O. Scully
Texas A&M University
College Station, TX 77843-4242
USA

Princeton University
Princeton, NJ 08544
USA

Library of Congress Card No.: applied for

British Library Cataloguing-in-Publication Data
A catalogue record for this book is available from the British Library.

Bibliographic information published by the Deutsche Nationalbibliothek
The Deutsche Nationalbibliothek lists this publication in the Deutsche Nationalbibliografie; detailed bibliographic data are available on the Internet at http://dnb.d-nb.de.

Cover Design Bluesea Design, Vancouver Island BC
Typesetting Asco Typesetters, Hong Kong
Printing and Binding Fabulous Printers Pte Ltd, Singapore

Printed in Singapore
Printed on acid-free paper

ISBN 978-3-527-40983-9

We are pleased to dedicate this book to

George P. and Cynthia Woods Mitchell

Modern Pythagoreans and patrons of science.

Contents

The Demon and the Quantum, Second Edition. Robert J. Scully and Marlan O. Scully

ISBN 978-3-527-40983-9

James Clerk Maxwell, Scottish physicist, *c*1850s.

> "In Science, it is when we take some interest in the great discoverers and their lives that it becomes endurable, and only when we begin to trace the development of ideas that it becomes fascinating."
>
> James Clerk Maxwell, *c*1850

Introduction

You can understand quantum mechanics if I can. I am a diesel mechanic. I grew up on a ranch 50 miles and 100 years southeast of Albuquerque. I attended college, became disillusioned, joined the Marines, and returned to civilian life as a mechanic and later an occasional biographer. Mostly, I have "learned" by the Oppenheimer method, as he said of himself:

> "I probably learned a good deal by a method that is never given enough credit, that is, by being with people … I should have learned

The Demon and the Quantum, Second Edition. Robert J. Scully and Marlan O. Scully

ISBN 978-3-527-40983-9

more mathematics. I think I would have enjoyed it, but it was a part of my impatience that I was careless with it."

R. J. Oppenheimer

I learned from my physicist father who, when I was growing up, worked at many places simultaneously: Los Alamos, University of New Mexico, Max Planck Institut für Quantenoptik, etc. We had a steady stream of famous scientists visiting our ranch, including Nobel Prize winners like Roy Glauber, Willis Lamb, and Julian Schwinger. My father was also a rancher, and I often argued with him and his partner Reese Woodling, President of the International Brangus Breeders Association:

"Why do you castrate calves and then give them shots to replace the growth hormones lost to castration? That's stupid."

I lost that one, because cut calves are more docile than young bulls.

"Why don't you explain engine operation using only one atom as the driving gas?"

I won that one, as we show in Chapter 3; it makes the subject simple and crystal clear!

Anyway, bit by bit, I coaxed a clear picture (of the quantum) out of them. What emerges is a curious blend of thermodynamics, quantum mechanics, and philosophy that I think you, dear reader, will like.

We often hear phrases like "quantum weirdness" and "the strange world of the quantum." What is not so widely advertised is that quantum mechanics can (and does) shed light on problems like the Maxwell demon paradox of thermodynamics, the seemingly "spiritual" nature of information, and even, perhaps, new insights into the existence of "mind" and (some say) the ways of God! The common denominator of all this is the fact that information is a real physical quantity. Information is more than just something in our mind or a databank. It is the essence of (and in a sense derived from) the study of entropy. Our theme is that thermodynamics, information science, and quantum physics are closely related. By focusing on entropy, information, and observation, we gain insight into the strange ways of the quantum. In fact, we might well frame the theme of this book in the form of the question:

> "Why can't we explain the recent developments mentioned above, e.g., resolving the Maxwell demon problem via quantum mechanics, without compromising the subject or patronizing the reader?"

We can – and that is the subject of this book.

My dad became sufficiently interested in the project to prepare a companion set of more technical notes (part of which is included in the Endnotes) for use in his lectures at Princeton, Texas A&M, and elsewhere. Hence, "proofs" of the various key ideas and results will be found in the Endnotes. Having explained the background and rationale for this book, let us sketch its contents and approach.

The notion that our thoughts and our mind represent part of a greater reality, as real as the things we think about goes back to Plato. Such a strange – very strange – idea has much in common with modern quantum mechanics. Only what we observe is real says the quantum physicist. John Wheeler teaches us that:

> "No phenomenon is a physical phenomenon until it is an observed phenomenon."
>
> John Wheeler

Werner Heisenberg and Erwin Schrödinger also underscore the central role of observation, perception, and information in modern science:

> "[S]eparation of the observer from the phenomenon to be observed is no longer possible."
>
> Werner Heisenberg

> "[Plato] was the first to envisage the idea of timeless existence and to emphasize it – against reason – as a reality, more [real] than our actual experience . . ."
>
> Erwin Schrödinger

James Clerk Maxwell introduced into thermodynamics "a being whose faculties are so sharpened that he can follow every molecule in its course." This being is "Maxwell's demon" who could, apparently, extract useful work from a single thermal heat source in violation of the second law of thermodynamics. Thus, the idea that consciousness or intelligence can be intertwined with physical phenomenon in strange ways dates back to 19th century thermodynamics.

Such an incorporation of intelligence into thermodynamics has profound implications. For the better part of a century, people argued about and puzzled over the Maxwell demon paradox. Finally, Leo Szilárd made the essential step of connecting entropy to observation and information in order to explain how it is that the "demon" does not violate thermodynamic wisdom. Walter Thirring, who is famous for his pioneering work in mathematical physics, comments on Szilárd's insights:

> "It is a pleasure for me to comment on one of Szilárd's scientific ideas, i.e., trading in entropy with knowledge. This seemingly crazy idea of equating something material with something purely spiritual turned out to be very fruitful."
>
> Walter Thirring

"Wait a minute," the honest, non-scientist might well say. "Before we go on any further, explain entropy to me. Sure, I always hear that entropy is a measure of disorder; the more clutter, the more entropy I guess. But what does that have to do with thermodynamics, engines, demons, and information?"

Answering that question in a simple, no nonsense fashion is the mission of the first part of this book, especially Chapters 3, 4, and 5. There is essentially no math needed to properly explain and understand entropy. To this end, we will focus on simple engines to explain the ideas of the great thermodynamicists, Carnot and Clausius, Maxwell and Boltzmann, and answer the question(s) of our honest reader. If you want to dig deeper, we provide detailed proofs (without using calculus) in the Endnotes.

After making the connection between engines and entropy (in Chapter 4), we then go on in Chapter 5 to introduce Maxwell's friendly demon and Szilárd's (one-atom) engine, thus making the connection between entropy and information.

Next, we go into the foundations of quantum mechanics and explore the experiments of Stern and Gerlach (SG) in Chapter 6. They used the magnetism of individual atoms to provide a royal road into, and demonstrate the need for, quantum mechanics. The Stern–Gerlach experiment also provides another window on the Maxwell demon problem. Following this line of thought, we arrive at a natural resolution of the demon paradox, without ever invoking the concepts of information and entropy, in Chapter 7.

Then in Chapter 8, we follow another road: the wave–particle duality route, into basic quantum mechanics. This approach yields Heisenberg's uncertainty relation, and sheds light on the role of observation and the observer in the quantum world. Maxwell would certainly have been interested to see how an "intelligent being" makes an appearance in another area of physics. In particular, in Chapter 9, we will explain how the acquisition of information and its subsequent erasure, commonly called "quantum erasure," can profoundly influence how we think about and understand certain aspects of "physical reality."

Finally, in Chapter 10 we will summarize various points of view concerning the ways in which quantum physics has been argued as affecting the ultimate questions of "mind" and even theology. The theologian Paul Tillich has observed that, among scientists, physicists are the only ones who seem to be able to talk of God without embarrassment. C. S. Lewis attributes this to the fact that physics is the most fundamental science. In addition, physicists are in a unique position to appreciate how little they know, and how natural the concept of God is in view of what they do know.

Telling all of this in a readable, coherent fashion is a tall order. However, I find that it helps if I first read the personal history of the scientists whose complex ideas are being studied. In this way one gains insight into their objectives and philosophies, thus into their discoveries and contributions. It is gratifying to see that the great Maxwell felt the same way, as per the quote at the beginning of this introduction:

> "In Science, it is when we take some interest in the great discoverers and their lives that it becomes endurable, and only when we begin to trace the development of ideas that it becomes fascinating."
>
> James Clerk Maxwell

The science in this book has a human side; the philosophy and goals of those responsible for the discoveries explored here are presented as well. This is certainly the case with artists, religious leaders, and public officials; their goals and beliefs are signposts pointing to their achievements.

Science is much the same. Who developed a certain theory and why? Why did Tom do such-and-such experiment and what did he tell Dick and Harry? To quote Leon Cohen: "who did what, when, and why." To answer such questions often goes a long way toward helping us understand their science.

After studying the scientist, we then proceed with a more in-depth study of the science. In this way, we first focus on the big picture – the whole forest – not the trees. Then, we proceed to dig deeper and think harder, often with the help of "thought experiments." In this way we gain a clear picture of the science involved. This is all accomplished without mathematics.

However, the reader who wants a more complete "proof" of the ideas and insights gained, may turn to the Endnotes where simple mathematics is used to sharpen our understanding. This allows us to develop the ideas and key concepts presented in the main text without encumbering that presentation with mathematics. The mathematical treatment in the Endnotes is presented in a self-contained form such that the necessary math tools are developed as needed. No math background is assumed. Further reading suggestions and connection with other presentations is also included in the Endnotes.

1

Mathematics, Mysticism, and More

From Pythagoras and Plato to Pauli

"All is number."

Pythagoras

"I confess, that very different from you, I do find sometimes scientific inspiration in mysticism ... but this is counterbalanced by an immediate sense for mathematics."

Pauli, in letter to Bohr

The unreasonable efficiency of mathematics in science is a gift we neither understand nor deserve.

Wigner

Wolfgang Pauli (Nobel Prize 1945) kicks a ball toward the camera of Roy Glauber (Nobel Prize 2005).

The Demon and the Quantum, Second Edition. Robert J. Scully and Marlan O. Scully

ISBN 978-3-527-40983-9

1.1 The Pythagoreans[1)]

Young Pythagoras was hungry, cold, and discouraged after his third night at the temple gate. His teacher Thales had taught him geometry and then sent him off to Egypt. He had been warned that entrance into the temple was no small matter, and it was not, for he had been turned away at temple after temple. At least here, they had not turned him away. "Wait." That is all the priest had said, his tone having as much emotion as did the yellowish desert sand that surrounded him. Through three days of blood-boiling heat and three nights of bone-chilling cold, he was tormented by blowing sand, its gritty texture wearing away his hope and spirits.

Just as he was about to give up, Pythagoras was finally invited to enter the temple. There, at the hands of his Egyptian masters, he learned math and mysticism, from the years 535 to 525 BC. After a decade or so, he was captured by the king of Persia, Cambyses II, who invaded Egypt around 525 BC. Iamblichus writes that:

> "[Pythagoras] was transported [to Persia] by the followers of Cambyses as a prisoner of war. Whilst he was there he gladly associated with the Magoi ... and was instructed in their sacred arts and learnt about a very mystical worship of the gods."

Some seven years later finds Pythagoras in southern Italy in the city of Croton, where he founded his famous school. There he created what is perhaps the first system of higher education that left a permanent mark on history. The philosophical–theological school of Pythagoras was unique. The story of his community has survived the eroding effects of two and a half millennia. As with many historical records, this one lacks color and vitality, yet the Pythagoreans offer an accurate and detailed account of the procedures governing their approach to higher education. Students accepted by Pythagoras took a vow of loyalty binding them to him and to their fellow students. The selection process was rigorous and demanding, much as the one Pythagoras himself had experienced in Egypt.

Many aspects of the Pythagorean school can still be seen in today's universities. His educational model had been molded by what he had experienced in Egypt, in his time the highest center of learning in the world. Before being admitted to the Egyptian institution, he first had to be examined by the elders. When accepted, he then had to pass an initiation, of which little is known. Thus, we have some of the most obvious earmarks of elite education as it persists to this day: highly selective, the desirability (real or perceived) of traveling abroad to get a broader education, as well as secret initiation practices. To clarify the term "secret" we note such initiation practices still survive in various fraternities/sororities today. Secrecy equals exclusivity, which in turn tends to generate interest and even some aspects of respect.

Because he completed his studies with the Egyptians and Persians, Pythagoras was respected and valued by society. Upon his return to Greece he had little difficulty in establishing a type of college loosely modeled on the one he himself had attended. More than just a college, it was also a community, in some respects similar to a monastery. Members' food and dress were simple and their discipline severe. One joined the Pythagorean community for life by surrendering all earthly possessions and totally submitting to the new system. It must have been worth the sacrifice, for there is no record of dissention among the initiates. The Pythagoreans were a self-supporting group and relied upon a system of communal property. They were also one of the first elite groups to recognize women as equals.

Pythagoras was the first to realize and to teach that proof proceeds from assumptions. The theorem bearing his name (see Fig. 1.1) may or may not have been his discovery, but he was clearly the first to introduce the idea of "proof" into mathematics.

The Pythagoreans were fascinated by the relation of number to musical tone. They determined the tone of a musical instrument was governed by the length of its strings. Decreasing the length of the strings or increasing the tension increases the pitch. This gives rise to the explanation of the physics of sound waves, another beautiful example of nature described by mathematical rules (Fig. 1.2).

The physics that applies to sound waves is similar to the physics employed in other wave phenomena. The waves have a frequency, measured in hertz, which is the number of up and down wave pulsations per second. The human ear is sensitive to frequencies between 20 and 20,000 hertz. The Greeks developed notes carrying six different frequencies on the handmade flutes they played (today's musical instruments employ 12 notes). The Pythagoreans thus described frequencies with the use of simple ratios. This inter-

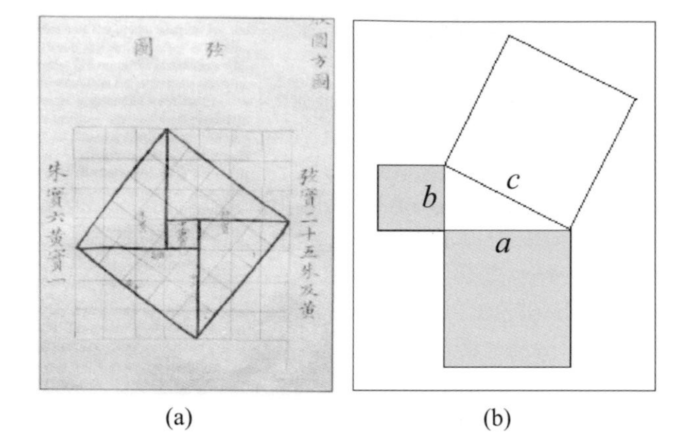

Fig. 1.1 (a) Chinese and (b) Greek approaches to the famous Pythagorean theorem, $c^2 = a^2 + b^2$.

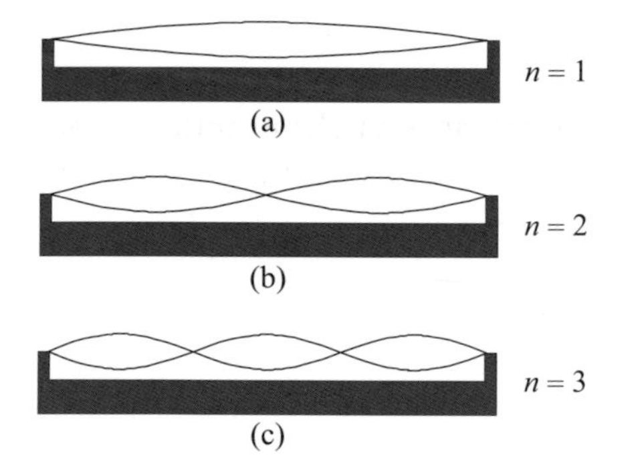

Fig. 1.2 The harmonies of a monochord. When a taut string (called a monochord) is plucked we see, and hear, first a note corresponding to vibration as a whole. This situation is denoted in (a). Then it vibrates in two parts, then three parts, as in (b) and (c), and so on. In this way, the tonal harmonies are seen to be described by numbers (a) $\to$ 1, (b) $\to$ 2, (c) $\to$ 3, etc.

esting interplay between math and music had a profound impact on the Pythagoreans.

The concept of math as an overarching theme intertwined with science, astronomy, and engineering is understandable, but why music? As the Pythagoreans proved, music *is* mathematical, and subjects with a mathematical foundation seem to become more and more analytical over the years.[2] The necessary mathematical rhythm in music is met when the chords involve frequencies that are multiples of each other. For example, if we play a chord at 440 hertz, then the other chords which will harmonize with it are those with frequencies of 880 hertz, 1320 hertz and 1760 hertz. But if we instead used chords of 600 hertz, 900 hertz and 2000 hertz, the harmonics would be lost and the listeners would leave the theater.[3] Science has yet to fully explain why the human ear enjoys this mathematical mixing of harmonics.

In an interesting parallel, radio antennas, as electromagnetic receptors, bear a striking similarity to the music preferred by our ears. If an antenna is broken, it will receive certain frequencies better if you break it at half its original length. These are obviously ratios necessary to produce harmonics like those used in music.

From music to astronomy, the Pythagoreans saw numbers as the answer to all questions. For this and other reasons, the Pythagoreans were fascinated with the intriguing realities involved in mathematical computations of any kind. In the harmony of mathematics they felt they had discovered an underlying form of consciousness and intelligence; in short, they thought they had found a path to god. More correctly, they had discovered their own chosen vision of god. Pythagoras was intrigued by such questions. His statement summarizes his philosophy:

"Reason is immortal, all else mortal."

The clearest, deepest form of reason is mathematics, and numbers are at the heart of mathematics. Therefore, it is understandable that Pythagoras felt that "God was number." At least he felt that "number" was the best path to his god.

"All is number" represents a philosophical reckoning of the universe by the Pythagoreans. In context, the early Greeks believed that the universe was made of four elements: air, earth, fire, and water. A similar attitude is voiced by the ancient Chinese, who believed that five elements made up the whole universe: metal, wood, water, fire, and earth.

The mathematical discoveries of the Pythagoreans were as logical as anything we are developing today. Parts of their philosophy were misguided, but many of their contributions are lasting. In fact, their discoveries are often what are taught today. Their analytical approach to geometry, their conviction that number is the key to the nature of the universe, and their belief in a communal life in service of religion are original Pythagorean concepts, ideas that still attract modern adherents.

There is at least one newer argument for the idea of "God" as number. IBM physicist Rolf Landauer made the point that information is just as real as electricity, water, rocks, and the wind. In this book, we will investigate this idea from several different points of view. We will reinforce the idea that information is a fundamental, "real" quantity, encountered in thermodynamics, computer science, and quantum mechanics. Even without such demonstration, we can see a connection between information and number.

Let us consider the state of a computer as nothing other than a bunch of numbers. Every element of "yes" or "no" memory is given by a simple number. Let us agree that the number is "1" for "yes" and "0" for "no." Mathematicians call this a binary number. At any given moment, we specify the state of the computer by a very large array of numbers, positive ones or zeros.

It is not a large step from there to think of our brain as a computer, and the eminent John von Neumann made this connection in his writings. The same argument would say that we are, at any instant, representable as a number. Such thinking once was strictly within the domain of philosophers, yet as we uncover more of nature's secrets, it is becoming more and more a part of working science. Perhaps the crazy Pythagorean notion of nature-as-number is easier to fathom today than in earlier times.

1.2 The Mind Expanders -1, $\sqrt{-1}$ and Ψ

"Minus times minus is plus for reasons we will not discuss."
W. H. Auden

Not surprisingly, mathematics, the science we use most in our daily lives, was the first science to be developed. All ancient societies developed mathematics. They began by using hieroglyphics and ingenious sequences of coded dots and dashes to facilitate highly adept forms of computation. These systems were used to aid in the construction of the massive building projects of antiquity's great cultures. The ability to pull off such engineering marvels at very early stages of cultural development shows much about the mathematical abilities innate to man.

The wonders of the world were often the mother of invention in the development of math. The ancient Babylonians relied upon a series of irrigation canals requiring many workers who needed to be paid, thus requiring a system of payment and record keeping. The Egyptians and the Mayans were very clever in the application of mathematics in the construction of their pyramids. During their golden age, the Indians explored the mysteries of π and contributed much to the understanding of algebra.

We are forced to use our imagination to comprehend their accomplishments, for too few records have survived to document the development of math. Yet, the history of math is a case study of the most frequently employed applications of pure science known to man. This history shows us a great deal about where we have come from and where we are bound. Fortunately, something of the story of math survived. For instance, we know the ancients were intrigued by π,[4] an irrational number (i.e., a number which cannot be expressed as the ratio of two whole numbers), just as they were fascinated by the circle and its intimate relation to π. They were somewhat philosophical in their thinking, emphasizing that the circle itself resembled π in having no stopping point. People in China, India, and Japan were particularly determined to compute this riddle number. In 264 AD, one Liu Hui constructed a polygon of 3072 sides and from this model calculated π to be 3.14159.

Unlike the number systems of other societies, which have been lost, forgotten, rediscovered, and rewritten, the Chinese have always used a decimal system in mathematics, and discovered the concept of zero early on. Indeed, the use of zero and negative numbers marks a watershed in mathematics. In today's credit card, debt-ridden society, negative numbers are not in the least abstract. Being broke (zero cash) is one thing. Being in the hole with a big debt hanging over our head imposes a painful dramatization of a negative quantity. When we are in debt, we have to pump cash into the system just to get back to broke. Negative numbers can be painfully real.

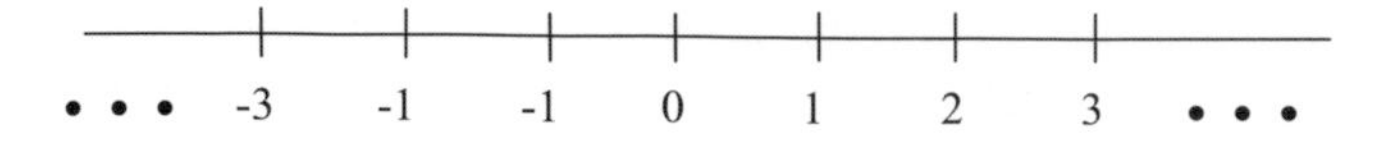

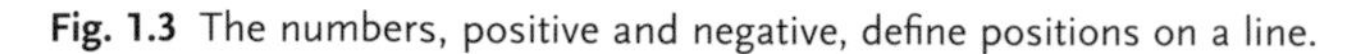

Fig. 1.3 The numbers, positive and negative, define positions on a line.

Another example of a negative quantity can be perceived by traveling in the wrong direction from one's destination. Thus, our velocity along the highway could be said to be +60 mph when we are moving toward our destination. However, if we make a mistake and are going in the wrong direction, then that is properly denoted as −60 mph.

In general, we draw a line and call the numbers to the right of zero positive. Those to the left are negative, as in Fig. 1.3.

As per the quote at the beginning of this section, minus times minus is plus. We perhaps recall this from our high school algebra. But why does minus times minus produce plus? What is the deal? Think of it this way: Suppose we are headed home at 20 mph and we double the velocity to 40 mph; this is just 2×20 mph. So far so good; now let's suppose we are going home at 20 mph and we flip a U-turn. Now, our velocity is clearly −20 mph, which is $(-1) \times 20$ mph. Then, if we flip another U-turn, our velocity is now $(-1) \times (-1) \times 20$ mph, and since we are clearly headed home again, we must be going +20 mph. Therefore, $(-1) \times (-1) \times 20$ mph $= +20$ mph, so:

> "Minus times minus is plus for reasons clear to all of us."
>
> R. J. Scully

Meanwhile, back at Pythagoras Polytech, things are not going so well. They have just learned that $\sqrt{2}$ is a strange (irrational) number, not expressible as the ratio of whole numbers. No ratio of two integers, no matter how complicated, will give the square root of 2. This alarming discovery really shook the faith of the Pythagoreans. If God is number, then how can a simple thing like $\sqrt{2}$ be imperfect?

But the square root of 2 is "imperfect" (irrational) and, consternated by this realization, Pythagoras's followers were pledged to secrecy. One of them (by the name of Hippasus, according to tradition) talked too much about the dirty $\sqrt{2}$ secret and, as the story goes, was drowned by his colleagues.

Just imagine if they had found out about negative numbers! More to the point, what if they had learned about $\sqrt{-1}$? Not only is there no rational number for $\sqrt{-1}$, there is no real number which multiplied by itself will give -1. In fact, $\sqrt{-1}$ is called "i," which stands for imaginary. This would really have had the Pythagoreans ready to mutiny.

But fast forward a couple of millennia to the time of the quantum. In quantum mechanics we use $\sqrt{-1}$ to represent nature. You can hear the Pythagoreans say, "This is just surreal." It is too mystical even for them! That is what science is all about – one of the messages of this book is that, however weird the mystics were, however strange their view of nature, the real world is even stranger than they (or we) would ever have imagined.

1.3 Let *x* Be the Great Unknown

"Al-jebr: reuniting broken bones."

Tradition has it that the Arabs brought algebra, alchemy (the forerunner of chemistry), and alfalfa to the world as they made their push westward. "Al" is Arabic for the definite article "the," thus alfalfa is "the grass." The word alchemy comes from the combination of Greek theory and Egyptian practical knowledge of materials, e.g., in mummification. This early fusion of Egyptian and Greek science was called "khemia" and the Arabs called it "alkhemia" or "alchemy."

Therefore, I always assumed that algebra stood for something like "the calculation" or "the arithmetic plus logic." Not so. According to the dictionary, the Arabic root is "Al-jebr," referring to the process of reuniting, as in the setting of bones. So the word "algebra" actually means something like "the method of unification," as in uniting or setting things equal. Anyway, those who claim algebra broke their bones will be interested to know that algebra is really more like resetting broken bones.

A little algebra makes a big difference in our ability to solve problems in modern science and technology. Algebra gets results quickly and accurately. According to my childhood encyclopedia, the modern-day high-school freshman knows more algebra than the wisest of the ancients 4000 years ago.

Simple algebra is the next math milestone after arithmetic and geometry. Arithmetic has proved indispensably useful, algebra even more so. A fair comparison would be using our fingers to dig in the garden (arithmetic) compared with using a shovel (algebra). For these reasons, we will present (and use) a little algebra in the Endnotes. Time spent there will be richly rewarded – try it, you will like it (and use it).

In particular, the use of exponential notation and logarithms is useful and important. For example, the number 100 can be written as 10×10, i.e., 10 squared, which we can write as 10^2. Likewise 1000 is 10^3 and one million is 10^6, etc. Any number can be written as 10 to some power or number. The actual power is called the logarithm of that number. For example, the logarithm of $100 = 10^2$ is 2 and the logarithm of 10^6 is 6, etc. Let me raise a flag here. The ideas of exponents and logarithms are very important. If these ideas are new to you, do not worry; they are simple – make them your friends. To learn more, please turn to Endnotes 1.

1.4 Math, Mind and Mankind

There are many similarities in the development of math among different cultures. Throughout time, we observe among various peoples insights into the workings of the human mind, and we gain thereby testimony to the mind's aptitude for working with numbers and geometrical quantities. How that aptitude has been developed is parallel to the manner in which societies developed. Unlike other forms of history, scientific history has the advantage that what is true and time-tested tends to be generally adopted, unlike political history, where the victor writes the history. Furthermore, the creators of science, as such, are competent. The facts always (eventually) win out. Nothing is more nearly perfect than math; no account of scientific discovery gains the day until reduced to number. A brief overview of math and number reveals the obvious as well as encourages more subtle inferences:

1. Math ultimately wins a hearing despite differing ideologies. As we have noted, the Pythagoreans tried to suppress the fact that $\sqrt{2}$ is an irrational number, but failed.
2. Early civilizations required some form of math in order to foster technological development, e.g., to keep track of inventories. Yet the connection between math and the mind goes deeper. Math begins to glimmer even in newborns. As we will discuss later, it would seem that humanity is born hardwired for math.
3. Societies on the rise develop and propagate math. Societies on the decline neglect education in general and mathematics in particular.

The last two statements call for further discussion. One might argue that this math–mind kinship affords evolutionary advantages. Successful hunting and protection of the clan require cooperation, organization, and logical planning. Mathematical thinking greatly assists activities vital to survival. Research into the nature of man and mathematical proficiency has been reported by Brian Butterworth[5] in his book *What Counts.* Butterworth found that mathematics might have been in use for tens of thousands of years, since he observed that cave drawings and Neanderthal relics often show the hash marks of the dot–dash method that was early man's usual form of numerical notation. He comments on research into the reactions of newborns to dot and dash configurations. Thinking patterns have been inferred in newborns by making observations of their stares and glances when shown pictures of objects. When shown cards displaying a series of dots and dashes, the duration of the infants' stares can be predicted by the number of dots and dashes with which the babies are confronted. Twice as many dots produce a stare twice the duration. Not that twice as many dots are twice as interesting. Rather, this evidence of infantile "counting" provides evidence for the innate mathematical orientation of human beings. Butterworth focuses also upon neurological studies of math prowess, a dauntingly complex, but nonetheless verifiable, phenomenon apparently inseparable from human nature.

Scientists like Butterworth are realizing that the teaching of numerical problems should be reformed to hasten and sharpen understanding. Introducing math to young minds can be substantially facilitated when the naturally hardwired talents of the human mind are taken into consideration. At present some members of the teaching establishment believe the calculator is a poor substitute for mental exercise; they argue that the device results in our declining to a nation of math illiterates who don't even know the multiplication tables.

We are currently repeating an old failure. As our schools decline, math is usually the first subject to be neglected. Consequently, the average American teenager displays a mathematical proficiency much below that of their Chinese counterparts. Collateral damage follows, with neglect of opportunities for the individual to develop reasoning and judgment skills. You cannot fast-talk your way to success in mathematics – answers are right or wrong. This pattern of neglect is not new, for throughout history math has been developing and advancing only later to be neglected for centuries. The past few decades of Western education have yielded mathematical backsliding.

After a few hundred years of advancing in mathematics, we are now beginning to slip. In the past, numerous obstacles hampered mathematics in the Western world. This continued until approximately the beginning of the 17th century. For example, measurements were imposed by the writs of kings, resulting in many absurd measurement systems that still linger in some regions. France, for example, had 2000 different units of measure in 1788. There was no uniformity of measurement from one city to another. This diversity was unnecessary. As much as one thousand years earlier far fewer and more reasonable standard measures had been adopted by King Charlemagne. Regardless of its other accomplishments, society can never become fully effective in science and technology if it lacks a solid mathematical base.

Is it possible that today we are underrating the importance of the first science discovered? If so, we should learn from the lessons of such negligence in the past. As noted by Beckman, the following words of the ancient Egyptian scribe, Ahmes, offer appropriate guidance as to what the foundation of true science is:

"Accurate reckoning. The entrance into the knowledge of all existing things and all obscure secrets."

What then has the work of the Pythagoreans brought us? It is another new piece of reality. Number is as real as air, earth, fire, and water, even though it is abstract. It is *very* useful and, as we will see in subsequent chapters, can be applied to what is real and knowable.

1.5 Plato – Another Kind of Pythagorean

Another tangible legacy of the Pythagorean school is its impact on later Greek scholars such as Plato. Indeed, the thoughts and insights of Plato deeply influenced quantum pioneers Heisenberg, Planck, and Schrödinger. According to biographer Armin Hermann[6)]

"In the course of his life, Heisenberg assimilated a great many intellectual ideas, and the greatest influence of all was platonic philosophy."

So who was this man Plato? Among other things, he was an extension of the Pythagorean theme. However, he was more of a philosopher than a Pythagorean. To him, "ideas" were the most real and valued entities in the universe.

Plato clearly learned from and was influenced by the teachings of the Pythagorean school, but Plato did not hold their philosophy of "number as all." However, he certainly adopted their views on much else. Plato saw in shapes, forms, planets, music, even reality and existence, the constantly guiding force of mathematical reason. This would become the theme of his own school, with its entrance door marked with a sign,[7)]

"Let no one ignorant of mathematics enter here."

Like many other philosophers, Plato felt that mathematics was the finest training possible for the mind.

As we shall see in later chapters, quantum physics seems to hold some sympathy for the views of Plato. For example, in the dialog *Timaeus*, Plato makes an interesting observation, which is very much in harmony with modern physics:

> "If in a discussion of many matters ... we are not able to give perfectly exact and self-consistent accounts, do not be surprised: rather we would be content if we provide accounts that are second to none in probability."

The quantum mechanic could not agree more. Quantum mechanics is probabilistic. Yet Plato does not think nature is totally random or chaotic. On the contrary, he stresses the role of an overarching intelligence in the universe. Plato, like the Pythagoreans, envisioned a perfect, mathematically elegant, universe. The fact that we often use a mathematical approach in everyday science and technology deepens our appreciation for and debt to the Pythagoreans.

1.6 Pauli – a Latter Day Pythagorean

It is fascinating that the foundation of quantum physics, one of the newest and most advanced fields of science, has aspects in common with the philosophy of Plato. Indeed, Plato finds an ally in the physicist Wolfgang Pauli (1900–1958), often regarded as the conscience of quantum physics. Pauli was one of the fathers of quantum physics as well as being a significant contributor to subjects ranging from Einstein's theory of relativity to the periodic table. Pauli is also regarded as the junk-yard-dog of quantum mechanics.

He was himself a prime mover in the formulation and application of quantum theory. For example, the Pauli exclusion principle explains how electrons conspire to make up atoms with only one electron per state. Some 15 years later he deepened our understanding of the behavior of electrons by using more advanced quantum mechanics (i.e., the quantum theory of fields) to show the relation between the exclusion principle and the internal properties (spin) of the electron. This is vintage Pauli – always looking for better, deeper understanding of nature and her ways.

Pauli was born in 1900 in Vienna, where his father, Wolfgang Joseph Pauli (WJP), was a professor of chemistry. Although Pauli's father was from a Jewish family in Prague, he (WJP) changed his name and converted to Catholicism. Wolfgang Pauli was baptized into the Catholic church and the famous Ernst Mach, whom we shall meet again later, was his godfather. His outstanding promise as a budding mathematician became apparent in his early years, and his father hired special tutors to train him in mathematics. He attended the university of Munich, earning his Ph.D. in 1921 under the leading light, Arnold Sommerfeld.

It is interesting to note that, although Pauli was known for his sharp wit and brusque manner, he was particularly deferential to his thesis father, Sommerfeld. After getting his Ph.D., he went to work for Max Born and then Niels Bohr, but he always treated Sommerfeld as *the* "geheimrat" professor. Around 1924, he developed the Pauli exclusion principle, mentioned earlier, and obtained a teaching position at the University of Hamburg. In the late 1920s, he moved on to the ETH in Zurich, which is the MIT of Europe. Except for a few years during World War II, when he was, among other things, "dean" of theoretical physics at Princeton, he spent most of his life in Zurich.

Pauli had many students and assistants who later became famous, for example, Victor Weisskopf, who was to become one of the most beloved, leading figures in physics. Weisskopf describes Pauli as being without guile and without regard for persons. He treated everyone alike: "big shots and small shots" got the needle equally from Pauli. He was not personal or vindictive, he just disliked sloppy thinking and half-baked ideas. As a result he would often criticize people in public: "This is stupid, go home and work it out" was a typical Pauli seminar criticism.

He was famous for his one-liners. For example, when a colleague (Paul Scherrer) presented an idea to Pauli and exclaimed, "Isn't that simple?", Pauli responded: "Simple it is, but it is also wrong." Another time when the University of Rochester asked for a recommendation for Weisskopf as professor of physics, Pauli responded: "I have nothing against this man." Everyone knew that was very high praise coming from Pauli and Weisskopf got the job.

Later in his career Pauli became interested in philosophy and mysticism. He collaborated with the famous psychiatrist and psychotherapist Carl Jung. Pauli was one of his favorite patients, and Jung made an elaborate analysis of more than four hundred of

Pauli's dreams. The psychological researchers of the day were in a different world from that of the hard sciences. Basically, however, the physical and psychological scientists were both trying to gain an understanding of the role of consciousness in human behavior. Their viewpoint was not unlike Plato's philosophy of reality and existence. They also had interest in the baffling questions posed by the role of the observer and the observed in quantum physics. But modern physical science demands experiment and proof. So Pauli made public very little of his research in this field, choosing instead to communicate his ideas through letters to colleagues and working in person with Jung.

This was, in some ways, similar to his mainstream scientific career, in which he published little, choosing instead to advance science by writing down his new ideas in letters that he sent to others like Heisenberg and Schrödinger. He wrote many thousands of these and it is likely that many findings that should have been credited to him went instead to others. But, no one ever doubted his genius. As Max Born said, "Since the time when he was my assistant in Göttingen, I knew he was a genius comparable only to Einstein himself. As a scientist he was, perhaps, even greater than Einstein. But he was a completely different type of man, who, in my eyes, did not attain Einstein's greatness." By age 21 he had finished his Ph.D., publishing a 237 page article on Einstein's theory of relativity. Einstein himself had nothing but praise for the work and it is still one of the most frequently used references to relativity today.

He never lost interest in the interplay between observer and the observed in quantum mechanics. Rather, his interest grew stronger and more philosophical. Below is a quote from one of his letters to Niels Bohr written in 1955:

> "What really matters for me is ... the more active role of the observer in quantum physics ... According to quantum physics the observer has indeed a new relation to the physical events around him in comparison with the classical observer, who is merely a spectator."

In another quote from the same letter (a part of which is at the beginning of this chapter), he expounds upon his thoughts on mysticism:

"I confess, that very different from you, I do find sometimes scientific inspiration in *mysticism* (if you believe that I am in danger, please let me know), but this is counterbalanced by an immediate sense for mathematics. The result of both seems to be my kind of physics, whilst I consider epistemology merely as a logical comment to the application of mathematics in physics."

As can be inferred from this quote, Pauli found that the conventional, scientific approach to nature is only one possibility. There are other approaches, such as the intuitive approach, which seems to be the antithesis of science but often provides valuable hints to solving problems. In this sense, Pauli saw logic and intuition as being two different (but not contradictory) approaches to problems.

Pauli won the Nobel Prize for physics in 1945 for the exclusion principle. He spent most of his life in Zurich, with a six-year period in Princeton as professor of theoretical physics during World War II. He left when Austria was annexed by the Nazis but returned to Europe after the war, settling in Zurich. Pauli was awarded the Max Planck medal in 1958 and in that year was later hospitalized with pancreatic cancer. When his assistant came to visit, Pauli asked him if he had noticed that he was in room 137. Throughout his life he had sought an answer to why the dimensionless fundamental constant involving Planck's quantum of action, the charge on the electron, and the speed of light, has the value of almost exactly 1/137. This important number in physics brings together electricity, relativity, and quantum mechanics. In that Zurich hospital on December 15, 1958, Pauli died in room 137.

Key Points

- Algebra unites.
- Exponential notation is neat, e.g., one thousand $= 10^3$.
- Logarithms are also cool, e.g., $\log_{10}(10^3) = 3$.
- Plato was essentially a Pythagorean whose ideas seem closer to those of a modern quantum physicist like Pauli than you would have ever guessed.

2
Mass in Motion

Kepler, Newton, and Kelvin lay the foundations

Tycho Brahe (1546–1601)

Johannes Kepler (1571–1630)

Brahe and Kepler combine measurement and mathematics to explain how planets orbit the sun.

"[Plato's] refutation of atheism turns on the identification of the soul with the 'movement which can move itself.' Thus, all motion throughout the universe is ultimately initiated by the souls. It is then inferred from the regular character of the great cosmic motions and their systematic unity that the souls that originate them form a hierarchy with a best soul, God, at their head."

Encyclopedia Britannica

The Demon and the Quantum, Second Edition. Robert J. Scully and Marlan O. Scully

ISBN 978-3-527-40983-9

"Plato conceived of a timeless eternity and I think that is something for philosophers to reexplore. For only in this way, I suspect, will we begin to reconcile our fleeting human existence in this particular universe with the larger cosmological structures that the incredible self-conscious brain of *Homo sapiens* can conceive."

Owen Gingerich[8)]

2.1 Measurement + Mathematics = Astronomy

I loved sleeping outside as a kid. The star-filled heavens above our New Mexico ranch were a wonder. There were arcing, shooting stars; glowing dull red stars; tiny piercing stars; glaring white stars; and distant, pinpoint galaxies, their photons struggling to reach out and communicate existence from what might be untold light-years away. I loved lying there wondering what this was all about. Observing all this as a child led me into that beautiful frame of mind ... curiosity, with a touch of wonder and awe.

Originally having found the orderliness of the stars to be scientific evidence for the existence of God, the ancients have much in common with the heaven watchers of our day. Astronomers search for cosmic order and discuss the God question with vigor. They tend to agree that for the universe to have happened the way it did by chance would require a miraculous fine-tuning. As discussed in Chapter 10, it would be like hitting a 1 mm bull's-eye from clear across the universe. The odds of a habitable biosphere, such as the one we have, simply seem to defy an explanation that involves freak chance or accident. The open-minded thinker (who does not worry about tenure or reputation) may still claim to see God's handiwork in the heavens.

Modern experimental science began to develop in the late 1500s. The order of the scientific community had begun to alter from a dispersed scene to an ever more international interaction facilitated by effective communication. Today this process has accelerated greatly, with entirely new fields springing up in the last few years (it has been estimated that scientific knowledge worldwide doubles every five years). A man who played a major role in bringing about this change was a brilliant Dane who separated superstition from science and through extensive observation and record keeping demonstrated the value of careful experimentation in the field of scientific discovery.

Tycho Brahe (1546–1601) answered to all the conventional expectations of European aristocracy. Handsome, strong, and articulate, he was also energetic, urbane, and affable. Known to duel on minimal provocation, he lost a big piece of his nose in one set-to and wore a copper prosthetic, held in place with some sort of ointment, for the rest of his life. He was nonetheless a gentleman who quickly forgave his opponents and could become lifelong friends with them.[9] Brahe was educated as a lawyer but began an affair with astronomy when he was 16 years old. He was also an expert in such other sciences as chemistry and medicine. His elixirs were of such high quality that they were sought after by kings throughout Europe.

Brahe was often called on by the Church to draw up ancient nativity charts, and frequently the Danish royalty (especially King Frederick) sought his astrological advice. Time-consuming as this was for Brahe, it was also beneficial to his career. The Danish King Frederick was so impressed with what he was getting that he gave Brahe his own island on which to build Europe's first large-scale research facility. It was a generously funded enterprise, absorbing about one percent of the Danish Crown's total yearly revenue.[10] Brahe's building skills would disappoint no one; he erected a mansion with a 16 furnace alchemical lab in addition to the observatory. The upper levels were supplied with running water from a spring-fed well (to date no one has determined how this was accomplished). An astronomer in every sense, he made the library circular. At its center was a six-foot brass globe on which over the course of many years he mapped the locations of one thousand stars.

He utilized numerous "high tech" instruments in his observatory, some of which took five or six artisans several years to build. Never satisfied with even the smallest imperfections, he was constantly removing some of them from service to be modified or rebuilt. He soon realized that an observatory built on wood beams was too wobbly and fragile, yielding inaccurate measurements. Therefore, he relocated his observatory outside onto a stone foundation with an observatory in which his instruments were each covered by conical roofs.

The technology and techniques developed by Tycho Brahe are impressive. He developed sights on instruments which eliminated parallax (an apparent change in the direction of an object, caused by a change in observational position that provides a new line of sight). He developed a new scale that gave accurate readings without simply enlarging the instruments to unwieldy proportions. He reversed the arrangement of the sextant to make a two-person instrument so that the planet and reference stars could be aligned simultaneously. He used multiple kinds of instruments simultaneously with a crew of observers to get altitude at the same time as angles between a planet and comparison stars. Brahe ingeniously used Venus as an intermediary to transfer solar coordinates to the night sky.[11)]

Tycho Brahe brought to science new technological methods of watching the skies. Data was compiled carefully over a period of years. Through his careful and painstaking data, he was able to prove that both the nova of 1572 and comet of 1577 were beyond the earth's atmosphere (contrary to the teachings of Aristotle). This helped pave the way for the collapse of Aristotelian physics. In the end, it was the sheer quantity of Brahe's work that carried the day. His many accurate astronomical observations added up to show an accurate picture of the heavens quite different from the one that formerly existed. Paradoxically, this quantity, in and of itself, constituted quality. It was careful, immaculate, and painstaking measurement that made the difference.[12)] Tycho Brahe was truly a premier empirical scientist. After Brahe, man needed to develop new technology (e.g., the telescope) to improve on his data. Indeed, it can be argued – and I do so argue – that modern measurement-based science began with Brahe.

Future developments based on Brahe's data would take teamwork. This was provided by Kepler's abilities as a mathematician. In particular, Kepler analyzed Brahe's data on Mars to show that Mars traveled around the sun in an elliptical (not circular) orbit. This was a very important discovery, since it was thought that the circle was perfect and every motion in the heavens was "perfect." It is to Kepler's credit that he found and reported this new "imperfect" elliptical motion. This compounding of intellectual ideas and skills was critical for the Brahe–Kepler success.

In any case, the famous laws of Kepler, describing planetary motion, should be credited to both Brahe and Kepler. Unfortunately, Brahe, the rightful co-owner of that fame, remains less well known.

Something of a Pythagoras of his era, Brahe's research heralded much of what was in store for the innovator. Plagiarism and intellectual expropriation of every sort were becoming (and would remain) constant threats to anyone with a new idea. Nearly every famous inventor thereafter would encounter idea thieves and battle intellectual bandits. Brahe fought the plagiarism battle for much of his life. For example, he even distrusted his own assistant, the (now) famous astronomer Kepler, in this regard. Their relationship began with a series of written correspondences.[13] This grew into a working relationship in which Kepler, according to some people, ultimately stole a lifetime's worth of data from Brahe. Brahe kept his work securely locked away; he had already learned his lesson. The two collaborated using Brahe's data and Kepler's mathematics, but it was not an easy relationship. Kepler needed Brahe's data to do his research but Brahe gave it to him only in dribs and drabs. Always determined to get his hands on Brahe's data, according to Joshua and Anne-Lee Gilder, Kepler poisoned him to get it.[14]

In pursuit of one of the world's oldest cold case files, the Gilders have shed new light on the subject of Brahe in their book *Heavenly Intrigue*. Brahe's body was exhumed and modern forensic science determined that he had been exposed to huge doses of mercury poisoning. So accurate was the autopsy, and so detailed are the records of the activities of Brahe's last days alive, that Kepler immediately became their number one suspect.[15] Determining exactly when the opportunity for poisoning existed, based upon calculations determining when the two were alone, causes them to point the finger at Kepler.

It should be noted, however, that not everyone buys the Gilders' theory. For example, the distinguished Harvard astronomer and historian Owen Gingerich says:[16]

> "About a year ago a book by the Gilders, a husband and wife team of journalists, proposed that Kepler had poisoned Tycho with a mercury potion in order to get his astronomical data. This would be like killing the goose that laid the golden egg, as Tycho was in the midst of negotiations with Emperor Rudolf II to have Kepler appointed Imperial

Mathematician, and with Tycho dead there would have been no guarantee that Kepler could wrest the observations from Tycho's very possessive heirs. The whole story simply doesn't ring true."

In fact it was Brahe, not Kepler, who was the knowledgeable alchemist. The use of mercury and its derivatives as an elixir or healing potion was widely practiced. It would seem reasonable to assume that Brahe himself prepared the mercury-laced brew that the Gilders refer to. Furthermore, murder would be out of character for Kepler. He was a very devout Lutheran, as his prayer near the end of his Harmonice mundi attests:[17]

> "If I have been allured into brashness by the wonderful beauty of thy works, or if I have loved my own glory among men, while advancing in work destined for thy glory, gently and mercifully pardon me: and finally, deign graciously to cause that these demonstrations may lead to thy glory and to the salvation of souls, and nowhere be an obstacle to that. Amen."

In any case, Brahe showed the world what it needed to know: science is best served by careful and well-planned measurements. Impeccable measurements and good data are the key to progress, and new technologies such as microscopes and precise chronometers would enable future advances. Nevertheless, the progress of science was still in first gear, and it would take years before Newton would surface to pick up where Brahe and Kepler left off.

2.2 Newton Invents Calculus, Revolutionizes Physics, and Minds the Mint

Based on the work of Brahe, Kepler, Galileo, and others, the man who really revolutionized science was, of course, Isaac Newton (1642–1726). His contributions were long-lasting achievements, in the same class as those of men such as Maxwell and Einstein. Like Brahe and most of the other "natural philosophers," Newton was a physicist, astronomer, mathematician, and alchemist, in that order. A leading figure of all time, he developed new science that proved central to all that followed.

Naturally, he did not accomplish this alone. Experimental science was coming to life at the time.[18] Great leaders and developers do not develop in a vacuum; they prosper when other geniuses pave the way. Newton, Maxwell, and Einstein all furthered the work of many others; they were in good positions to make key breakthroughs in their field.

It was during, or soon after, Newton's time that scientific societies like the Academia del Cimento (Academy of Experiment) and the Royal Society were being formed.[19] These associations provided encouragement and recognition for the leaders of science. Scientific periodicals also appeared in Europe during this time.[20] Papers were being published, ideas exchanged internationally, and collaboration on new ideas becoming commonplace. The new atmosphere provided for a technical revolution as the power of novel ideas began to focus and to influence occupations that are more specialized.

The passing centuries have worn the details, trials, and intellectual adventures of historical science into what is all too often taken for granted. With the old theories now long-standing laws, we shall not bother to look into the mathematics and research that brought them into existence. However, the mathematics and imagination that prompted the discoveries of Galileo and Newton are truly impressive. Their complex figures, calculations, and theorems ran for pages. The basis for their research looked little different on the surface from many of today's scientific efforts. Often the amount of scientific advance is commensurate with the amount of math used. Early scientists used advanced forms of algebra and geometry, even if the methods they used were less user friendly and of a different "language" than those employed today.

Newton himself invented calculus, which he used thoroughly in developing his theories of the gravitational attraction of one planet upon another. This advancement in science was astonishing. Like many major breakthroughs (the laser, thermodynamics, discoveries in astronomy, etc.), the authorship of the calculus was hotly contested. Unlike Brahe, Newton was not a dueler. This may have been fortunate for his adversary, Gottfried Leibniz, who according to historical record developed a form of calculus independent of Newton. Much controversy existed during their lifetimes over who developed this new form of mathematics, but enough documentation exists in Newton's favor to assure him his rightful seat as the father of calculus. Nevertheless, Leibniz made original, seminal contributions to the calculus independent of Newton.[21]

In his book *Principia*, Newton lays out much of his life's work. This includes his three laws of motion and his universal law of gravity. His theories (now "laws") were inspired by phenomena observed, not on earth, but in space. Sir Isaac solved mysteries that were several centuries old when he accounted for them. For example, Kepler's elliptical planetary orbits had already been discovered in the previous century, but it remained for Newton to show *why* planets move along elliptical orbits.

Newton's first law states that a body at rest will remain at rest unless a force acts on it, or that a body moving in a straight line will continue to move with the same speed in a straight line unless a force acts on it. (It had been previously thought that a body required a constant force to keep moving, because moving objects eventually always stopped without an external force pushing them.) Newton attributed this "stopping" to friction, something that would play a unique role in the coming explorations into the workings of energy.

Newton's second law says that the rate of change of momentum (momentum is defined as the mass of a body times its velocity) of a body is equal to the force applied. For example, an object in a constant gravitational field (as on the surface of the earth) feels a constant force; therefore it falls at an increasing rate of speed because the force acting on it is fixed.

The third law of motion is that action and reaction are equal and opposite. This law can be surprising when, for example, it tells us that the falling apple pulls on the earth just as the earth pulls the apple. The miniscule mass of the apple of course does not have much effect on the earth, but the notable point here is that the odd-shaped elliptical orbits of the planets were explained by this "pulling" law.

Newton's three laws of motion and his universal law of gravity revolutionized physics. Today, the three laws of Newton are the basis of classical mechanics. His laws remained unchallenged until Einstein's theory of relativity, in which space and time could be united by noting that the speed of light was a constant for all observers. Classical physics was not undone by relativity, but was not valid when speeds approach the speed of light.

It is an English understatement to call Newton a hero. It seemed he had the golden touch. Indeed, they even put him in charge of the mint. Nevertheless, he had less success with his theory of light. He did the right thing with prisms and telescope lenses, but when he attempted to explain the behavior of light he got it wrong. He rejected the wave theory of light.

Newton's research was always on target. He conducted his light experiments very close to the way they are carried out today. He shone a strong beam of light through a tiny slit and then watched the pattern it made on a wall. He tried to see if the beam would bend when it encountered a corner as a wave would. However, he found that light seemed always to travel in a straight line after it passed through a slit.

Were light like other wave phenomena (e.g., water waves), it would surely bend around corners. However, he could detect no such bending, and this led Newton to believe that light did not display wave-like behavior. He therefore concluded that a beam of light must consist of particles, like a stream of machine gun bullets. The problem was that existing technology was not yet advanced enough to study light effectively. The wavelength of light was too short to display wave-like behavior in his crude experiments. Ultimately, Newton was (sort of) right about light – it is in many ways particle-like. But it also has a wave side to its nature, something he did not suspect. Had he been equipped with better technology, he would have actually seen wave-like behavior.

Wave behavior would finally be shown by a physician and scientist named Thomas Young, who demonstrated the wave nature of light over a century later. Until then, the particle nature of light was accepted, largely due to Newton's stature.

Newton's laws and much of his research were attempts to understand the mysteries of the natural world. Most science of this time addressed astronomical issues, but it would not be so restricted for much longer. The research of a handful of men like Brahe, Galileo, and Newton had set in place the foundation for a new experimental approach to science and technology that would benefit all of humanity directly and soon change the way the world was run. Newton and Galileo established a new paradigm that natural laws were expressed in mathematical form. Beginning with Brahe, science had become a precise tool governed by experiment; henceforth all theories had to be tested and proved. Evidence, not conjecture, became the unofficial motto of research. Newton continued this tradition when he defined motion in ways that produced not just philosophical schools of thought but proven rules of nature and laws that determine the working limits of science. Newton united celestial and terrestrial mechanics. His was the first breakthrough discovery by which regularities observed in planetary motion provided models to explain the behavior of matter on earth. This reciprocity has become a familiar feature of scientific research: the distant often reveals the near-to-hand and vice versa.

2.3 Matter in Motion: Momentum and Energy

One other aspect of motion we need to flag is the kinetic energy of a particle. This is the energy a particle has due to the fact it is moving; and is governed by the mass of the particle times its velocity squared. Compare this to the momentum of a body, which equals its mass times its velocity. There is an important correlation between heat and energy of atoms in, for example, gas. In those days people did not know that heat, energy, and motion were all interrelated; a few intuitive individuals with their subtle but subsequently famous observations went a long way toward developing the intellectual material for an impending scientific revolution.

To take a step back chronologically, Francis Bacon[22)] (1561–1626), the great evangelist of science, said:[23)]

> "Heat is a motion, expansive, restrained, and acting in its strife upon the smaller particles of bodies."

Bacon lived more than a century before Newton, but his statements were insightful glimpses into the ties among motion, heat, and energy. Subtly and cleverly are these phenomena interwoven, yet their connections are remarkably logical, even obvious after the fact. Bacon made a few insightful statements pointed in the right direction without quite arriving at the destined terminus. What was needed was a more precise method of measurement in order to gain a deeper understanding of what energy and motion really are.

Sir Benjamin Thompson, Count Rumford (1753–1814), was an interesting player to the music of Bacon and Newton. He understood heat in the same way they had, and carried out measurements that would explain away much of the confusion. During his time, a primitive belief still prevailed regarding the existence of heat. Heat was thought to be an invisible substance called "caloric." Caloric was believed to be contained in all matter. It was somehow released during burning or when two objects were rubbed vigorously together. When something ceased burning because it exhausted all combustible fuel, it was thought that all the caloric had been burned off.

Strange as it may now seem to us that such a belief would persist so long, it definitely would have to be superseded if thermal physics were ever to really get underway. Rumford took note of the caloric fallacy when he was in charge of a cannon-making factory near Munich. Having an investigative temperament, he paid close attention to the manner in which heat was released during the boring of cannon. The cannons were solid metal prior to being subjected to extensive boring. After a huge horse-drawn drill bit had bored for a certain length of time the metal became hot enough to boil water. As an experiment, Rumford submerged a cannon completely in water and drilled until the water boiled. He then stopped boring and when the water had cooled, he began boring again. According to conventional theory, the "caloric" (or at least some of it) should have already been removed. But, it always took the same amount of time for the water to begin boiling when the boring was resumed. This was contrary to caloric theory, according to which all heating of the water would cease when the caloric was used up.

Rumford almost discovered the first law of thermodynamics. In an offhand remark, he stated that the water surrounding the cannon could be brought to boil more efficiently simply by burning the horses' hay under the container of water, rather than feeding it to the horses, which then turned the drill bit. He unwittingly made an educated guess toward the energy/industrial revolution's first step: discovering and utilizing the principle of the conservation of energy. We note the important connection between energy and motion of a body. The energy of a moving body goes as its velocity squared; this is explained further in Endnotes 2.

Rumford was a hundred years ahead of his time and did not realize just how precisely he had hit on the right track. His ideas about heat were not immediately seized upon. It would be several generations before the laws of thermodynamics, implicit in Rumford's statement, would resurface in forms ever more polished and accurate. So many scientific phenomena defy our basic instincts and everyday perceptions of reality and existence. This counter-intuitive character of science has led scientists to say that sometimes our minds are our worst enemies.

Examples of these "unreal realities" are available in those sciences that run contrary to our intuition or expectations. Atomic physics and quantum mechanics afford probably the most well known example. This counter-intuitive trend began with thermodynamics, and then progressed into an understanding of entropy, which in turn permitted the development of theory explaining atoms (heat carriers, if you will). The scientists of the 1800s were originally very troubled over the idea of atoms, as some believed that the existence of matter at its smallest level would somehow compromise "proofs" for the existence of God.

Nevertheless, the scientists who grew up in the 1900s proved the existence and developed the understanding of atoms, as they are known today. It is interesting, however, that the quantum pioneer Eugene Wigner recalls[24)] his high school text as saying that: "atoms may or may not exist; it makes no difference to us in our study of physics." However, to Wigner and his friends, atoms *were* very real. They developed the new science of quantum mechanics. Among other things, they proved that matter and energy could change forms, thus allowing for the advent of nuclear technology. Information control has done much more to transform lives, leading to the simple (if unnerving) conclusion that science and technology, and their usefulness for humanity, are often proportional, not to their conformity with current understanding, but to their non-conformity. It seems like, the stranger and more bizarre the scientific mystery, the more beneficial the resulting applications have proved to be. For example, quantum mechanics, with all its subtleties, led to the transistor, with all its utilities.

New scientific innovations have always branched off slowly from the proven foundations. Knowing how those original concepts were established and having a clear understanding of them is the key to grasping later subjects that are more complicated. A brief look at the discoveries made by the "founding fathers" of energy theory reveals strong parallels between the concepts of energy and heat behavior and the behavior of the phenomena of light, consciousness, and existence discussed in this book.

William Thomson, Lord Kelvin (1824–1907), was one of a handful who facilitated the critical steps in the right direction. As had Rumford, he too, realized that heat needed to be better understood and defined. To that end, Kelvin introduced a perfect temperature, known as absolute zero, the complete absence of heat. It is important to note that the average energy of atoms in a gas is determined by its temperature. In fact, one degree absolute (called one degree kelvin, or 1 K) is essentially the energy that the atoms have at that (1 K) temperature. Absolute zero (zero degree kelvin, or 0 K) is a perfect temperature in that you can add more heat, but you can never take any more of it away once you get down to 0 K (which corresponds to −273 °C).[25] Cold is simply the *absence* of atomic and molecular motion: you cannot add cold to any substance (no matter what its temperature), you can only take heat away from it. You do not feel the cold of an ice cube in your hand; you feel the uncomfortable absence of heat in your hand as heat travels into the ice cube. That is, you "pump" the heat out when lowering the temperature of something. An advance was made; the difference between hot and cold temperatures involves energy.

Kelvin was not the only visionary to realize that understanding heat was going to be the key to unlocking new technology. The industrial age would advance only as fast as the development of new ways to control energy permitted. Brahe, Newton, and Kelvin were all giants who were moving discovery and understanding toward a similar juncture in science and technology, but much more needed to be known. There were secrets to uncover; modern science was just getting underway.

Another visionary, James Joule (1818–1889) was a close friend of Kelvin, with similar interests. He sought the secrets of energy in the form it takes in hydropower, which at the time was a preferred source of power in England. An expert at detecting the minutest changes on a thermometer, Joule made his mark on the world by proving that the movement of water creates friction. He even measured the minute temperature differences between the water at the top and bottom of a waterfall. In this way, he showed that the energy of falling water could be transformed into heat. Like Kelvin, he knew that more accurate forms of energy *measurement* were required if we were ever to learn enough about energy to progress beyond the ungainly looking, inefficient, and unreliable steam engines of the day. Today, we honor Joule by calling the basic measure of energy the "joule."

Following Galileo, Brahe, and Newton, the greats of discovery focused on measurement. Whether it was the area of triangles, the temperature of water, the motion of molecules, or the concepts of motion on earth or in space, all were matters requiring measurement. To measure is to begin to understand nature. Understanding nature is the key to the control of nature.

Today, it is more of the same. From new forms of energy, to unlocking the DNA code, scientists are attempting to measure and fathom the secrets of nature. Although the inventors are still busy trying to make their next million, they seldom build anything in advance of the technological secrets revealed. That is why the harnessing of energy is so difficult for the inventors in the transportation businesses to improve upon. Gasoline is widely used because it burns easily. The worldwide proliferation of gasoline-driven cars and trucks, and the resulting fuel shortage and pollution, are undesirable but inescapable consequences of that fuel choice. And such must be the case until our scientists and engineers gain further control of nature, which fortunately is entirely plausible (e.g., controlled fusion). If we understood the secrets of nature, would any endeavor be beyond our reach?

That last philosophical question would not likely find a consensus among scientists. The discovery of previously unknown phenomena has always opened the doors for much philosophical thinking. It often happens that those involved in new discoveries become deeply philosophical, and some even religious, due to insights gained by their research. Kelvin and Maxwell were staunch Christians, while others like Einstein spoke often of "God" or "The Old One." This deity was the force responsible for controlling the mechanical perfection of the universe, a force that very few believe could operate so perfectly by chance. This is a junction where the scientists' search for the mechanical definition of life and nature coincides with the philosophers' moral and human studies of life and nurture.

The philosopher often seeks to find the meaning of life and existence through riddles and questions such as those asked by Erwin Schrödinger:

1. Does an "I" exist?
2. Does a "world" exist besides "me"?
3. Do "I" cease to exist on bodily death?
4. Does the "World" cease to exist on "my" bodily death?

The question "What is real?" is always at the forefront of the philosopher's mind, as well as "What is consciousness and existence?" The scientist, too, cannot help but find himself confronted eventually by such thought. As we uncover more of the deep secrets about math and science, it is beginning to become obvious that *these* are indeed the fabric of reality. Consciousness is the understanding of that reality. Existence is prerequisite for reality and consciousness. The creator, God, controls all and is all. When we learn science we learn what "all" is and how it is controlled. The understanding of "all" is achieved through measurement, which will eventually tell us the meaning of the whole.

In the words of Victor Weisskopf:

> "Nature in the mind of man recognizes herself."

Key Points

- Momentum goes as velocity.
- Energy goes as velocity squared.
- Temperature is average energy.

3
From Engines to Entropy

Carnot and Clausius develop the yardstick of engine efficiency

Rudolf Clausius (1822–1888)

Sadi Carnot (1796–1832)

Carnot unveils the second law of thermodynamics and Clausius introduces the concept of entropy.

"Carnot's theorem is the most impressive example in all of physics of the triumph of theory over praxis, of abstraction over experience. It stands as a monument to the value of pure science – not just to the intellectual endeavor of understanding nature, but directly to the practice of engineering. Thus it establishes science as the very heart of industry and commerce."

Hans von Baeyer[26]

The Demon and the Quantum, Second Edition. Robert J. Scully and Marlan O. Scully

ISBN 978-3-527-40983-9

3.1 Sadi Carnot (1796–1832)

The angry cries and horrified screams overlay the rhythmic clatter of the horses' hooves. The drunken cavalryman was indiscriminately lashing out at pedestrians with his saber. So drunk he could barely stay on his galloping horse, he had accidentally done serious injury to several of those he meant only to frighten. This was supposed to be beautiful, tranquil Paris, but one often gets back what one sends out. France had been "sending out" and waging wars quite regularly during the early 1800s. Debt, defeat, and revolution were the results of the actions of the Bourbon Dynasty and Napoleon. And deranged soldiers. Where had this one been demoralized? Was it the freezing, grueling defeat in Russia? Or in Prussia? Then there was England and Sweden, not to mention the revolution, as well as the many outposts, which of course were deemed to require a constant military presence. These were to be found in far flung places wherever there was a profit to be made. In those days of colonial power, might was right. Today represented another loss to the superpower. The drunken cavalryman was surely a disappointment to himself and his army. Wherever he came from, whatever lost campaigns he had participated in, he was a disgrace to the cavalry, traditionally the elite of any military.

Another horrified scream as his saber slipped and sheared away a piece of a woman's scalp. How could he take pleasure in this? Was not France a land with honorable traditions and noble gentlemen? It was, and one man who was a reminder of this gentility stood in the crowd of panicking civilians. Coolly and expertly that man stood to one side of the galloping beast and grabbed its marauding rider just as he sped past. The maneuver was so sly the horseman never knew what hit him. Our hero unseated the drunkard and threw him into the gutter.

Fortunately, our man was up to the task of unhorsing the berserk trooper; he himself was a highly trained officer in the best tradition of the French Army. He was short and slight, but as an experienced swordsman he had developed wrists and forearms the size of a much larger man. Good with a saber and super-agile, he was adept at the martial arts. One might expect no less from the son of one of Napoleon's top generals. He was also an engineer and a brilliant scientist. Sadi Carnot was patriotic, and his intentions were to see his country take its place as a world power alongside (and perhaps even ahead of) England. France's revolution was a success, but Frenchmen had not fared well afterward in the wars with European monarchists.

The world in general, and France in particular, could have learned much by carefully listening to Carnot, but it was not to be. Possibly because of pride and arrogance, the scientific community missed Carnot's point; it was due to the fact that Carnot's presentation was not directed just to the scientific community. It wasn't that Carnot didn't develop his ideas without employing careful math and physics. He declined, however, to rely on mathematical proofs in publicizing his thoughts. He wanted to reach everyone, not just scientists and engineers. The result was that his brilliant ideas were not appreciated by a scientific community that did not think his work was "serious" research. To add to his difficulties, even purged of mathematics, the layman didn't understand Carnot's explanations.

Carnot was a scientist, but one equipped with military prowess and instinct. He looked to France's long-time rival and enemy, England, to determine which areas needed strengthening at home. Carnot attributed England's success in the industrial revolution to her use of steam power. He even claimed that, if England were to lose her steam power, the effect would be more dramatic than if she had lost her Navy. We propose to follow Carnot's path, and in this section we will show why every heat engine requires a source of both hot and cold, and then provide an account of Carnot's treatment of engine efficiency. In particular, we will present a simple toy (single-atom) Carnot engine, a model which allows us to understand how engine efficiency depends on the way we add and remove heat. This will lead into the introduction of the entropy concept of Clausius in the next section, and that of Boltzmann in the following chapter.

Carnot, the military strategist, engineer, and scientist, formed a plan. What was the enemy doing? Few today would question the role of energy and power in an industrial society, but in the 18th century most of the world did not truly understand the value of becoming an industrial society, let alone the mechanics that provided the heart and soul of such an economy. France had steam locomotives, but they were primitive machines with engines which lumbered along at about six percent efficiency. Something needed to change. Some obvious improvements were to be made in engineering and design features, but, as we are learning today, *substantial* improvements to any industry result from technological and scientific changes, with engineering producing thereafter a physical means of implementing those changes. Difficult as it might be to believe, scientists and engineers of the time could build engines, but they didn't understand the physics of heat nor the reality of energy. Only later would they learn the first law of thermodynamics, namely that heat added to work done on a system equals its change in internal energy.

Carnot, the son of Napoleon's Minister of War, showed early on signs of courage and integrity. When he was four, he chastised Napoleon for splashing water on a boat full of women, shouting, "You beastly First Consul, stop teasing the ladies!" As part of the aristocracy, the little fireball got an education in math, science, language, and music. He cut short his study of engineering in order to join the Army. He refused any of the privileges he might have enjoyed on account of his famous father and delved into his hobbies of literature, the arts, and music with all his effort. From his determination to excel, one would expect Carnot to go far, and, indeed, the fields in which he surpassed his contemporaries were well chosen. His was one of the best minds at a time when society was just taking on its modern character of constant mechanical innovation. Hindsight prompts us to say that his interest in the steam engine was right on target. Few other Frenchmen in the early 1800s had any interest in steam engines, perhaps explaining why at the time these primitive devices were as inefficient as we have observed they were. When Carnot sought the reason for such massive waste, he discovered that poor design was only partly to blame. Heat was produced by burning coal, but a totally unappreciated fact was that none of the heat could be used unless *some* of it could escape. Today this is obvious. Every power stroke of an engine (i.e., firing of fuel in the cylinder) is accompanied by an exhaust stroke which throws away or wastes energy. Unfortunately for Carnot's compatriots, the science and engineering of that time could not determine this necessity from monitoring the fierce flow of heat as it coursed through the steam engines. The necessary waste of heat energy on the exhaust stroke of an engine is just a fact of nature. Observation of, and deductions from, this simple fact are the basis of the *second law of thermodynamics.*

The second law of thermodynamics states that heat flows only from hot to cold, and never the other way around. Energy in the form of heat follows the path of least resistance, flowing only down the entropy hill, so to say. The second law of thermodynamics teaches us that you can't get useful work from a single heat reservoir, e.g., from running an engine without an exhaust stroke. According to the second law, trying to use the heat in the ocean to run a ship without doing anything other than cooling the ocean would be a futile undertaking. Like the alchemists who tried for millennia to convert lead to gold, many attempted to find a way to violate the second law. But they couldn't (and can't); that law always decrees the working limits of engine operation. More on this later, but if this principle did not exist, extraction of useful work from the heat stored in any system, such as the ocean, would be possible on an essentially unlimited scale. The so-called perpetual motion machine (of the second kind) would be a reality. We say "of the second kind" because it would violate the second law of thermodynamics. A perpetual motion machine of the first kind would violate the first law of thermodynamics, i.e., the conservation of energy.

Modern engineers actively employ Carnot's ideas and in doing so build better engines. Furthermore, they have realized the benefit of heeding the cautions Carnot issued in order to know what will and what won't work. Thanks to Carnot's researches, we are now designing industrial systems that achieve maximal efficiency by being so planned that they scrupulously obey the laws of thermodynamics.

In his attempt to compete with England's success with steam power, Carnot sought to better understand the concept of heat/energy transfer in an engine. What limits the engine's efficiency? How can one get the most work out of a pound of coal? He understood that some of the heat put into the engine had to be removed in accord with what we might consider a novel application of the "what goes up, must come down" commonplace.

3.2 The Hot Rod Engine

Fire alone is not sufficient for mechanical payoff; it takes fire and "ice" to run an engine in an endless cycle of hot–cold, hot–cold, hot–cold. The "ice" is better known as an exhaust stroke in a fuel-operated engine. To give a simple example, consider the toy steel rod engine sketched in Fig. 3.1. The steel rod expands on heating and contracts on cooling, thereby propelling a mass up an inclined plane. In the first step the rod is heated and the upper mass moves upward one tooth on the inclined ratchet. Then the coolant (i.e., ice) is applied and the bottom mass is pulled up one tooth. This

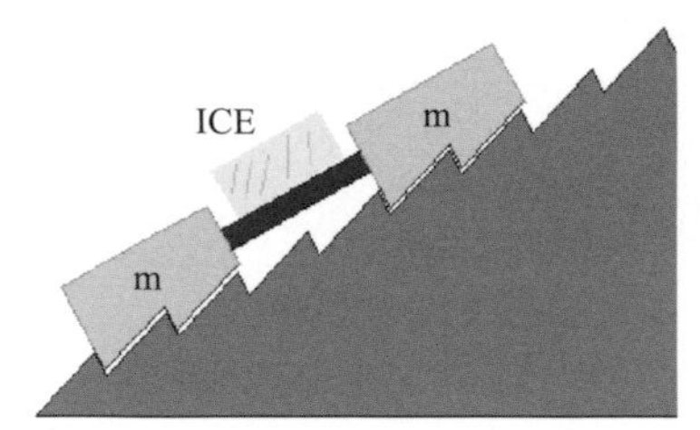

(a)
Start with ice-cold rod

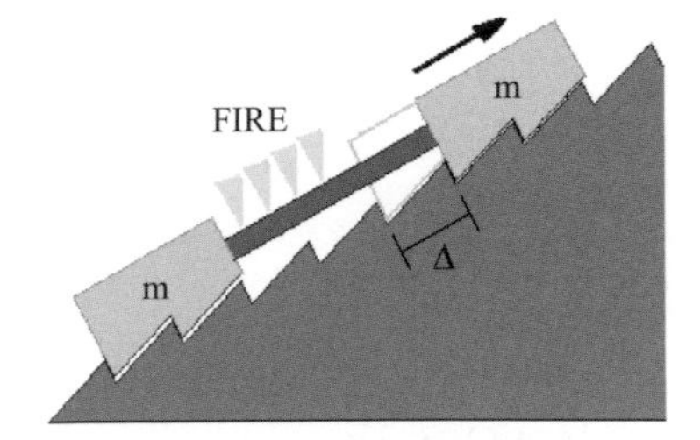

(b)
Heat rod, upper weight pushed up locks in place

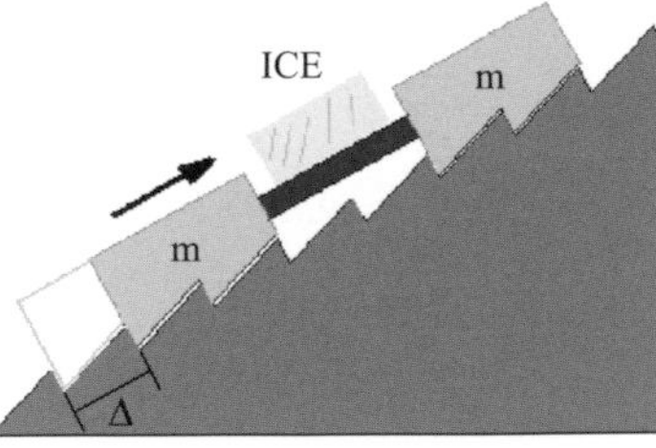

(c)
Cool rod, lower weight pulled up locks in place

(d)
Repeat heating and cooling process engine climbs hill

Fig. 3.1 The toy expanding rod engine. The steps of operation are as follows: (a) Start with an ice-cold rod. (b) Apply heat to the rod, expanding it, and pushing the top weight up the hill, while the ratchet teeth prevent the lower weight from slipping back. (c) Now cool the rod with ice, so it pulls the lower mass up. (d) Heat the rod again, so the top mass moves up the hill, repeating the cycle.

cycle of hot and cold is applied over and over to achieve useful work, i.e., to lift the pair of masses up the inclined plane.

The toy engine of Fig. 3.1 has much in common with an external steam or internal combustion engine in which a piston pushes down on a connecting rod to make an energy (or power) stroke, and then comes back up in an exhaust (or cooling) stroke. Heat in produces work out. But some of the heat is always wasted. How can we minimize the waste and derive the maximum work from a given amount of heat input?

In order to answer these questions, we must consider how we can define and thereby determine the efficiency of an engine. This was something Carnot was keen on, since he wanted maximum efficiency. The toy rod engine of Fig. 3.1 is useful in emphasizing the importance of two heat baths, one hot, one cold. It takes both "fire" and "ice" to actually run an engine. However, we will now replace it by another super-simple (single-atom) Carnot engine whose operation, especially as related to its efficiency, we can more easily understand.

3.3 The Carnot Cycle Engine

At this point, let's turn to the simple single-atom (Carnot) engine of Fig. 3.2, in which we replace the rod with a single atom and use the impact of the atom against a piston as the mechanism for powering our engine. We call this the *Carnot cycle engine* because we follow Sadi Carnot in the way we apply heat and take it away to make the best engine. The atom bounces back and forth between the piston and the cylinder head many times during one power stroke. In this way the atom will communicate its kinetic energy to the piston. (See Endnotes 3 for more detail.) Naturally, the atom will begin to lose energy and slow down as it does work on the piston unless we do something to supply more energy. This is where, once again, the heat comes in. If we apply heat to the walls of the cylinder, this enables us to keep the atom at the same temperature during the power stroke. This is called *isothermal expansion* and is pictured in Fig. 3.2(b). Next, we turn off the heat and allow the piston to further expand as per Fig. 3.2(c). The

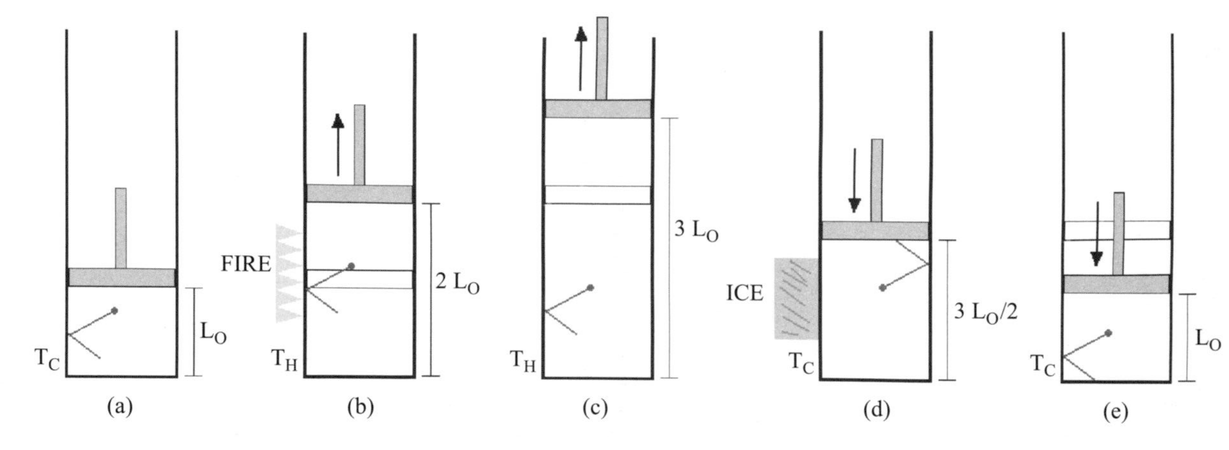

Fig. 3.2 This is a single-atom Carnot engine. In moving from (a) to (b) the single-atom volume expands, at constant temperature T_h, so that distance L_0 becomes distance $2L_0$. We then remove the heat source and allow the piston to continue to expand to $3L_0$, which takes us from (b) to (c), and completes the power stroke. In keeping with the laws of thermodynamics for any engine, we must now compress the single-atom gas *back* to its *initial* volume. We next compress the gas at T_c (which is indicated by the ice cube in (d)); this takes us from (c) to (d). That is, we compress our one-atom gas to $3L_0/2$. Lastly we remove the ice cube and compress the single-atom gas until we are back at the starting point; this takes us from (d) to (e).

a → b → c expansion produces the useful work, as is further discussed in the caption as well as Endnotes 3.

Continuing to follow the engine operation, we note that after the piston has reached its full extent (bottom dead center), we obviously must now think in terms of cranking it back to the original (top dead center) position. However, if we do this with our cylinder maintained at the original high temperature, we would have to do as much compression work on the atom as the atom gave to the piston in the a → b → c expansion. It is, therefore, clear that what we want to do, just as in the previous toy rod engine, is to use a source of cold (water or ice) to cool the atom, i.e., the cylinder walls, in order to carry out compression at a lower temperature. In particular, the energy associated with compression will now be proportional to a lower temperature, which we call T(cold).

The efficiency of this simple single-atom engine is defined as the energy consumed on the power stroke, minus the waste work of the compression stroke, all divided by the energy consumed. We show in Endnotes 3 that the efficiency of our ideal (Carnot) engine is determined solely by the temperatures of our hot energy source T(hot) and the colder reservoir T(cold). Nothing else!

Carnot had much to contribute, but it is as if that wasn't meant to be. Not only was the world unable and unwilling to learn from him, but his life was cut short. Carnot died of cholera in 1832, when he was just 36 years old.[27] Because France was gripped by an epidemic, his clothes and most of his research papers were burned. Fortunately, a small bundle of his notes was spared. Not published until 1966, they are a testament to Carnot's simple but brilliantly effective ability to interpret nature. His discoveries of the laws of thermodynamics have stood the test of time because they were based on a deep understanding of the nature of heat exchange and engine operation.

3.4 Clausius' Entropy

The German physicist Rudolf Clausius (1822–1888) took over where Carnot left off. It might be more nearly correct to say he rekindled the flames Carnot had ignited. Moreover, unlike Carnot, he made exhaustive efforts to insure that his new research would be well received and accepted by the scientific audience of his day.

The two men's lives and objectives were mirror images of each other in several revealing ways. In seeking the meaning of thermodynamics, Carnot and Clausius both worked along similar lines. In many ways, their work, goals, and mentalities were very simpatico. Working in parallel in complicated, and at the time, unexplored fields, the two men reached strikingly similar deductions.

Both men were fiercely patriotic: Carnot, devoted to strengthening his country against its external enemy, England; and the older Clausius, who volunteered to serve in the Franco-Prussian War of 1870–71. Though 50 years old, Clausius was a valiant commander, receiving the Iron Cross, and later becoming disabled in the line of duty. He was in charge of an ambulance corps that rescued injured soldiers from the battlefields (which corresponds to the career of Carnot if we reflect on his actions cited at the beginning of the chapter). His strong sense of patriotism was part of an emerging trend. Actually, his generation was the advance notice of a groundswell in ultra-strong nationalism. Prussia, to become Germany, would move toward centralization, eventually culminating in the fanatical state "worship" and diabolical schemes for concocting the perfected Reich envisioned by Adolf Hitler.

Clausius had numerous running battles with several famous foreign scientists whom, he insisted, could not be given credit for discovering thermodynamics. To his mind, such enlightenment could only come from German minds. Carnot, another nationalist, was also an eccentric. These two giants of science had much in common. Both men were sons of well-to-do, well-educated fathers. Both were excellent horsemen. They both sought the most comprehensible explanation for thermal phenomena, striving to present their work in terms that others could easily understand.

Clausius undeniably had Carnot beat on that last count. He even used descriptive, vivid terminology to good effect. He was a perfectionist who was unusually successful in describing and simplifying subtle concepts. Furthermore, Clausius wrote for his fellow scientists, unlike Carnot, who tried to write for the layman and scientist alike. Clausius' reflections on energy and heat engines led to deep insights into the nature of matter and even into the metaphysical imponderables of existence itself. He knew that the sum total of energy in the universe was constant. However, in order to describe how energy is used to drive an engine, he introduced a new quantity he called "entropy." Entropy governs the ability of a system to utilize heat energy to produce useful work; that is, it takes both a (hot) source of energy and a (cold) sink of heat to run an engine.

We emphasize that on the isothermal compression stroke entropy is removed from the atom, and added to the cooling water. Developing his theories further, Clausius stated that the entropy of the *universe* tends toward a maximum state of disorder. Let's follow Clausius to see how the study of engines and energy revealed a deep insight into all of nature. To see how this comes about, we return for a moment to Carnot and his heat engines (Fig. 3.3).

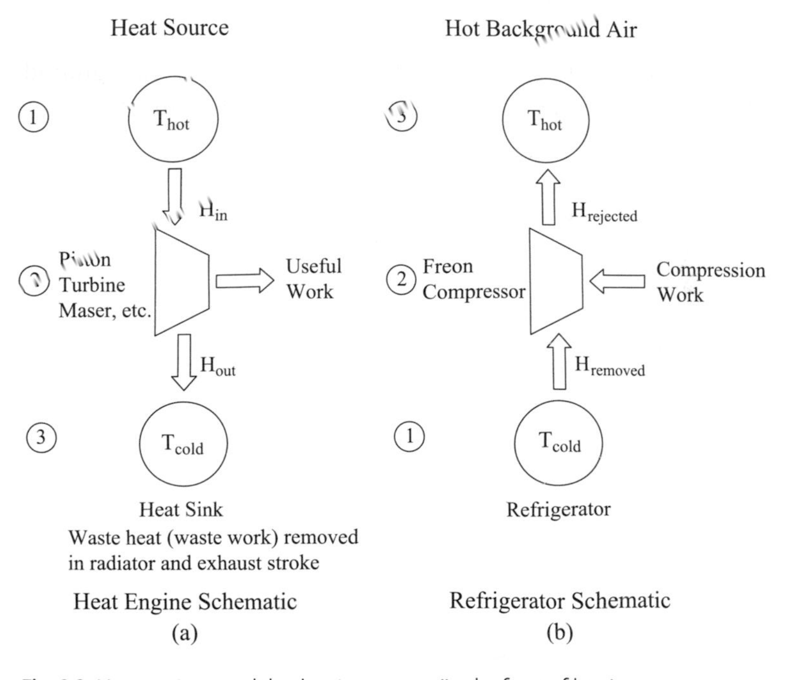

Fig. 3.3 Heat engines work by drawing energy (in the form of heat) from a high-temperature energy source, using the energy to power an engine, and rejecting heat to a lower-temperature heat sink. Air conditioners and refrigerators do just the opposite, i.e., are heat engines running backwards.

Carnot realized that all reversible (e.g., no friction, etc.) heat engines have the same efficiency. To see this, look at Fig. 3.3, where we depict a reversible heat engine operating between hot and cold reservoirs. Now suppose that it is possible to make an engine with greater efficiency. We can take the output of that engine and run the first engine backwards, thus yielding a net transfer of energy from cold to hot. But if we could do that, we could make a perpetual engine of the second kind. That we cannot do, as the (second) law of thermodynamics tells us. So we must conclude that all reversible heat engines have the same Carnot efficiency. In Carnot's own words:

> "Now if there existed any means of using heat preferable to those which we have employed, that is, if it were possible by any method whatever to make the caloric produce a quantity of motive power greater than we have made it produce by our first series of operations, it would suffice to divert a portion of the power in order by the method just indicated to make the caloric of the body B return to the body A from the refrigerator to the furnace. To restore the initial conditions, and thus to be ready to commence again an operation precisely similar to the former, and so on: this would be not only perpetual motion, but an unlimited creation of motive power without consumption either of caloric or of any other agent whatever. Such a creation is entirely contrary to ideas now accepted, to the laws of mechanics and of sound physics. It is inadmissible. We should then conclude that the maximum of motive power resulting from the employment of steam is also the maximum of the motive power realizable by any means whatsoever."

The Carnot result is so important that we should reiterate. It doesn't matter if we run our engine on coal, natural gas, or uranium nuclear fuel; all that matters is how hot you can heat the "boiler" and how cold you can make the cooling water (or ice, etc.). Furthermore, it doesn't matter what molecules (steam, air, Freon, etc.) you use to drive the piston. Indeed, it doesn't even matter if you reduce the number of molecules or atoms to *one*, as we have done; the engine has the same efficiency. What a fantastic result!

Carnot made his discoveries by focusing on engine efficiency. And, as mentioned earlier, entropy is the key concept in the study of thermodynamics. It was Clausius who gave us that insight. Only an outline of Clausius' ideas and discoveries will be given here. The entropy concept will be further explored in the next chapter and in Endnotes 3. For now, we will keep Carnot's ideas of the movement of heat in mind, as that is a big help in understanding entropy.

Clausius found that the heat taken in on the power stroke divided by the temperature of the hot reservoir equaled the heat wasted on the exhaust stroke divided by the temperature of the cold reservoir. This is proven in Endnotes 3. When something is constant or unchanging (like this heat-to-temperature ratio), the quantity is said to be *conserved*. In this case, we call the ratio of heat change to temperature "entropy." In our reversible operation, the entropy we added to the gas on the expansion stroke (at the higher temperature) is equal to the entropy which we then removed by the compression stroke (at the lower temperature). We have arrived at the notion of entropy in the context of the heat engine, but the concept has much broader application even to the study of "mind" and information.

The deeper insight of Clausius will be further illustrated in the next chapter. The point is that the entropy increase of the power stroke must be removed on the compression stroke. Entropy goes up on the expansion stroke at higher temperature. It comes down when we compress the gas at the lower temperature. We could say that what goes up, must come down, entropy-wise. This "conservation" of entropy only occurs in ideal reversible engines. Real engines increase the entropy of the universe on each cycle. The mechanically operated piston is always in a state of moving up and down, compressing and decompressing (expanding) the gas. On the expansion cycle of the piston, the entropy is increasing, only to be again decreased on the compression stroke.

Clausius started the entropy legacy with an acknowledgment that the sum total of the energy in the universe is constant. Remember the law of energy conservation, in which energy is neither created nor destroyed, but only changes form. However, Clausius added the new insight that the amount of entropy in the universe is ever increasing. In other words, entropy in the universe is always increasing. Why?

First, consider an infinite expanse of empty space (prior to the big bang). Second, think of the entirety of matter in the universe compressed down to the size of a dime. Third, imagine the big bang. Fourth, realize that, among other things, entropy has emerged. Entropy increases as all matter and energy continues to expand outward into the void of space.

A brief glance at the history of modern science will show entropy as being the next step in a series of necessarily sequential discoveries. To uncover the realities of science, you must first climb her ladder; every rung of discovery is close to, but slightly above, the last one. We started at the bottom, with our discovery of heat and energy. However, entropy represents the next rung on our ladder of understanding. In the following chapter, and indeed in the remainder of the book, we will further investigate the concept of entropy.

Key Points

- Carnot proved that ideal engine efficiency is determined only by the temperatures of the heat source and entropy sink.
- Clausius introduced the concept of entropy generated by a system as the heat generated divided by the temperature.

4
From Statistical Entropy to Statistical Time

Thermodynamics evolves from being an engineer's coal pit to becoming a philosopher's gold mine

Ludwig Boltzmann. (Photo courtesy of J. Bevan Ott.) (1844–1906)

Max Planck (1858–1947)

> "But forgive me, if before we go on I ask you for something that is most important to me: your confidence, your sympathy, your love, in a word the greatest thing you are able to give, yourself."
>
> Ludwig "($S = k \log W$)" Boltzmann

By studying the entropy of light, Planck is led to the quantum.

Here, for the first time in human history subjectivity, our state of knowledge, makes its debut in science. This subjective something (S) is called *entropy*.

The Demon and the Quantum, Second Edition. Robert J. Scully and Marlan O. Scully

ISBN 978-3-527-40983-9

4.1 Entropy and Statistics

It is to Maxwell and Boltzmann that we owe the deep insights and analyses of the kinetic theory of gases.[28] Before their mutually reinforcing work, the formulations of the principles described in the previous two chapters were foggy and often incorrect. For example, many men of science, including the famous German physicist, Max Planck, did not initially believe in atoms. Another example is the Austrian physicist Ernst Mach (think Mach 1, Mach 2, aircraft speed, that's the guy), who never believed in atoms. They did not agree with the picture of heat as being random molecular motion. Eventually, Planck did come to accept atoms as real and went on to convince not only himself but also the rest of the world that even energy comes in clumps or quanta. After Boltzmann had died, everyone accepted the key idea: Atoms do exist and the motion called heat is nothing more and nothing less than the random motion of atoms and molecules. This bit of wisdom is the cornerstone of modern science. As Feynman puts it:[29]

> Question:
> "If, in some cataclysm, all of scientific knowledge were to be destroyed and only one sentence passed on to the next generations of creatures, what statement would contain the most information in the fewest words?"
>
> Answer:
> "The atom is the fundamental structure of matter: All things are made of atoms – little particles that move around in perpetual motion, attracting each other when they are a little distance apart, but repelling upon being squeezed into one another."
>
> Richard P. Feynman

Based on his "heat equals atomic motion" picture, Ludwig Boltzmann (1844–1906) gave us a new, clear insight into the nature of entropy. He found that entropy is determined by the number of "places" or "states" an atom or atoms can occupy for a given fixed energy.[30]

If this sounds like a departure from the previous example of entropy, it's not. It's just a more detailed description of how energy and matter behave. This insight is clearly presented in Boltzmann's writings and lectures. In the chapter heading we only half-in-jest listed the famous equation $S = k \log W$ as his middle name. (Actually it was Planck who first wrote the entropy equation in this form, as is discussed briefly in Chapter 6.) Fittingly, that equation appears on his tombstone. This formula relates a system's entropy S to the logarithm of system probability W. The constant k is the Boltzmann constant we mentioned in the last chapter. This simple formula deserves to be as famous as Einstein's $E = mc^2$. Our first task in this chapter is to understand and make friends with this brain child of Boltzmann's.

We need only look to our single-atom Carnot engine to get an example of Boltzmann's equation at work. In Endnotes 4 we show that the entropy generated in our one-atom working fluid when we double the volume is given by $S = k \log 2$. The explanation for this is that, when we move the piston so as to double the space the atom can be in, the entropy has doubled. There are now two "places" for the atom to be, the top volume or the bottom volume. This change in entropy is governed by $k \log 2$, as Fig. 4.1 makes clear.

If, instead of doubling the volume, we tripled it, then we would have three equal places for the atom to be (note that the atom's energy is still the same), so $W = 3$.

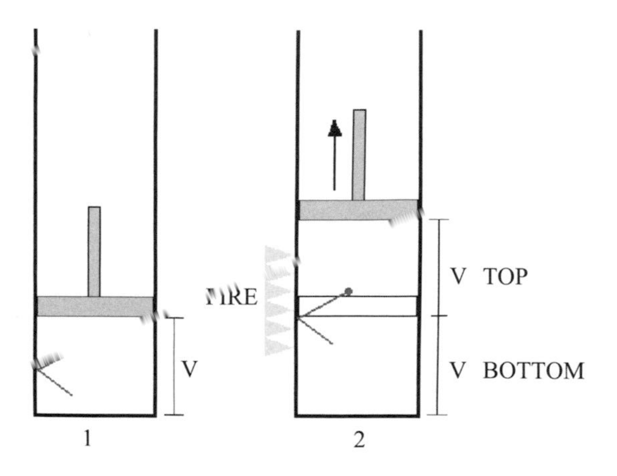

Fig. 4.1 Isothermal volume doubling. When the piston expands so as to double the gas volume isothermally, our atom can now be in the top or bottom volume. This means that W (i.e., the probability) is given by $W = 2$, indicating that the atom can be in one of two places.

More generally, W determines the entropy generated by the isothermal expansion of a gas as it goes from the small volume v to the large volume V. That is, $S = k \log W$ where W is the large volume V divided by the small volume v, i.e., $W = V/v$.

As another example, suppose our atoms have internal states they can occupy. Take the number of states to be two for simplicity. Call the more energetic state the excited state and the other one the ground state. Now there are again two places the atom can "be" and so we say the internal entropy is $k \log 2$. If we had N states, we would have entropy $k \log N$. The number of equivalent "places to be" or "configurations" determines W. So W is called the *configuration number* and $S = k \log W$ is called the *configuration entropy*. This is the heart of thermal physics.

Let's move beyond the entropy involved in the single-atom engine. The concepts of energy and entropy really come into their own in many-atom systems. Consider a pool table and balls. The balls prove to be a helpful model for imagining the individual molecules and atoms and their overall interaction with each other. The containing rails of the pool table are like the walls of the cell that contains our gas or liquid. For our purposes, a gas is the preferred state because it is much easier to identify the movement of individual atoms when they are spaced further apart as they are in a gas.

In Fig. 4.2 we show only two balls ("atoms") and constrain motion to be in the vertical direction. Suppose we arrange energy

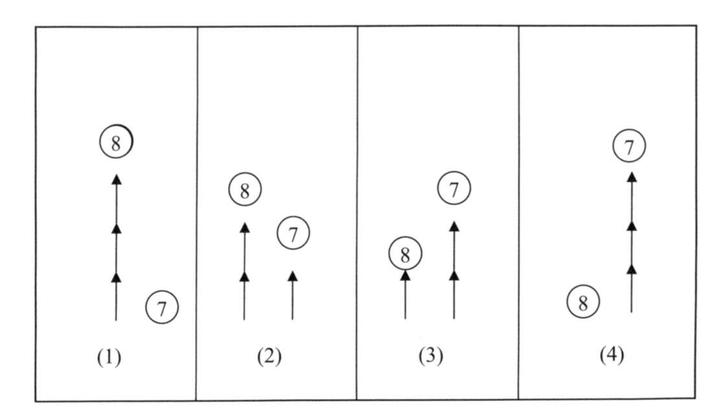

Fig. 4.2 Three units of energy distributed between balls 8 and 7 constrained to move in the vertical direction.

to come in lumps of amount E_0 so that each atom can have some number of such lumps. Further, suppose both atoms have three lumps of energy altogether. Then the number of configurations of energy $3E_0$ is $W = 4$, as shown in Fig. 4.2. For a gas of many atoms, the configuration number is huge.

4.2 Entropy and Information

"... we discover a remarkable likeness between information and entropy. This similarity was noticed long ago by L. Szilárd, in an old paper of 1929, which was the forerunner of the present theory. In this paper, Szilárd was really pioneering in the unknown territory which we are now exploring in all directions. He investigated the problem of Maxwell's demon, and this is one of the important subjects discussed in this book. The connection between information and entropy was rediscovered by C. Shannon in a different class of problems, and we devote many chapters to this comparison. We prove that information must be considered as a negative term in the entropy of a system; in short, information is negentropy."

Leon Brillouin[31)]

As noted by Brillouin in the preceding quotation, Szilárd made pioneering inroads into information science by combining his studies on entropy, information, and Maxwell's demon. The entire next chapter will be devoted to Szilárd's study of the demon problem.

In this section, however, we'll adopt Claude Shannon's approach to information and entropy. Following Shannon, we call the number of possible outcomes of a process N. For example, the number of possible outcomes of flipping a coin is $N = 2$ (heads or tails equals two outcomes). Likewise, the number of possible outcomes of throwing a die is $N = 6$. These outcomes represent information. In fact, they *are* a form of information. Then, following Shannon, we define the information I by the equation $I = \log N$. Note the close correspondence with Boltzmann's entropy $S = k \log W$.

Going back to our example of Fig. 4.1, the physical connection between information and entropy is clear. The more room the atom has available to be in, the less we know about where it is. If the atom is in a volume V and we double the volume to $2V$, then we have cut our knowledge of where the atom is by half. If we triple the volume we cut our knowledge by one-third, and $S = k \log 3$.

The combination of Shannon's information and Boltzmann's formula for entropy, like all deep truths in nature, is at bottom quite simple. Mathematically speaking, entropy is a measure of probability. It's like the machine you may have seen mixing up numbered ping-pong balls in the lottery drawing. Our formula for entropy expresses the system's state of disorder. On comparison, the balls in the lottery mixing machine are in a tremendous state of disorder and understanding entropy helps us to quantify the randomness of this or any other system. (By "disorder" we mean the literal opposite of order, in which all the balls would be neatly organized in neat rows with all the same numbers on them all facing the same direction. That's nearly impossible to achieve just by chance, which is why disorder is the way of nature.)

Entropy applies equally well to large systems like the weather, evolution, and the cosmos. It helps us to understand the likelihood of creation and the expansion of the universe with its mystifying grip on time. The problems involved in determining the interactions of many molecules are very much like those required for plotting the interactions of a correspondingly large number of stars. The molecules in our ocean, the clouds in our sky, and the many galaxies in the heavens all obey Boltzmann's law of entropy as surely as a falling apple obeys Newton's law of gravity.

Many people contend that the increasing disorder in the universe points to an initial creation event, relative to which entropy is constantly increasing. They argue that this did not come about purely by chance. It is the result of a higher, creative controlling power.

As pointed out in Chapter 3, the realities of nature are to be understood by learning how to measure those realities. Measurement is the capacity to reduce to number and to define those phenomena we know as reality. Information and entropy can describe randomness and order. The toss of the dice or mixing of the lottery balls are anything but orderly. Orderly would be if all the trees in the forest grew in perfect rows in perfect formation from shortest to tallest. The grains of sand in the desert, the waves in the ocean, and the clouds in the sky show both order and randomness. For anything but chaos to arise as a result of the big bang initial event requires certain special non-linear mathematical structures that we are only just beginning to understand.[32]

The developing field of information science is a hot topic in science today. And the measurement of information has now been correlated with entropy. Entropy increase has been in effect since the big bang. Beginning with the study of simple engines, we have discovered a deep truth. In that regard Boltzmann says:

> "The scientist asks not what are currently the most important questions, but 'which are at present solvable?' or sometimes merely, 'in which can we make some small but genuine advance?' As long as the alchemists merely sought the philosophers' stone and aimed at finding the art of making gold, all their endeavors were fruitless; it was only when people restricted themselves to seemingly less valuable questions that they created chemistry."

One might expect the men who gave us such undreamt-of insights to have been well received, but during their lifetimes they were often under-appreciated. Visionaries confront the problem that the rest of the world cannot see far enough ahead of existing technology and methods to recognize that the notions of some gifted minds are novel, yet true. Boltzmann found this problem to be trying and demoralizing.[33] Toward the end of his life, Boltzmann fought increasing bouts of depression. His health began failing, his eyesight becoming so poor that he could not always do physics. He relied more and more heavily on his assistants and family, a difficult thing for an independent thinker and leader. It became more than he could stand, and, on a summer day in 1906, Boltzmann took his own life in the hotel room where he was staying.

But the wonderful teachings and deep thoughts of Boltzmann live on. No better example of this exists than the beautiful creations of his disciple Erwin Schrödinger concerning statistical time.

4.3 Entropy, Statistical Time, and the Eternal Mind

> "The theologian Paul Tillich once remarked that among his professional colleagues, of those in the sciences, only the physicists seemed capable of using the word 'God' without embarrassment. Physicists have known this for a long time. Many examples come to mind, Einstein heading the list."
>
> Bryce De Witt[34]

One reason why physicists may be more likely than other scientists to believe in God is the fact that they understand what it is that they don't understand. C. S. Lewis, long-time Oxford professor, relates his experience with scientists along the following lines:[35] Biologists tend not to believe in God and when pushed to the limit admit that concerning the ultimate questions they don't know, but they say the chemists know the answer. Chemists likewise mostly don't believe in God and also don't know, but they contend that the physicists know. The physicists, however, know they don't know and they tend to be open-minded seekers.

One such "open-minded" (to say the least) scientist is Erwin Schrödinger.[36] He contends that "time" and more importantly "statistical (entropy) time" provides a window on the eternal soul! He reasons as follows:

> "The most important contributions from science to overcome the baffling questions 'Who are we really? Where have I come from and where am I going?' ... I say, the most appreciable help science has offered us in this is, in my view, the gradual idealization of time."

He goes on to pay homage to the ancients:

> "[Plato] was the first to envisage the idea of timeless existence and to emphasize it – against reason – as a reality, more real than our actual experience; this, he said, is but a shadow of the former, from which all experienced reality is borrowed. I am speaking of the theory of forms [or ideas]."

He then also discusses the statistical physics of Boltzmann who gives us the "arrow of time" concept. Schrödinger explains that the very notion of past and future comes from statistical reasoning:

> "To my view the 'statistical theory of time' has an even stronger bearing on the philosophy of time than the theory of relativity ... may, or so I believe, assert that physical theory in this present stage strongly suggests the indestructibility of Mind by Time."

The statistical time concept that entropy = time's arrow has deep and fascinating implications. It therefore behooves us to try to understand entropy ever more deeply. Entropy not only explains the arrow of time, it also explains its existence; it is "time". This was one of the first observations relating information (entropy) to our actual experience in nature. Clausius and Boltzmann were intrigued by this idea. As we noted earlier, time begins to tick and move forward as our system generates entropy. What does this mean? One way to better understand this idea is to return to our Carnot engine. There, we found that entropy of amount $k \log W$ (where W is the ratio of the final to initial volumes) is removed from the boiler (heat source) and added to the atom (working gas) which drives our engine on each power stroke. However, we remove this amount of entropy from the atom (transferring it to the cooling water) on the compression stroke.

Note that the Carnot engine operates very slowly in order to accomplish the exact cancellation of entropy added (to the working gas) on the power stroke and removed on the compression stroke. Real heat engines do not achieve the efficiency of Carnot's engine. Carnot's engine generates no entropy; the entropy taken from the boiler is equal to that dumped into the cooling water. Real engines do generate entropy and are not reversible. In general, the entropy of the world is always increasing. This is why scientists and engineers say entropy provides an arrow of time.

Hence we see that "time", as we have constructed it, is subjective, in as much as it is related to our experience, i.e., to our state of knowledge. This statistical time is not the time of Newton, which flowed independently of all events – his time was similar to the space intervals marked out by a ruler. Statistical time is different. It is connected to the second law of thermodynamics and is thus not something we are embedded in or carried along by. Sure, our bodies age, but our thoughts, and our mind (as opposed to our brain), according to Schrödinger, do not. Henry Margenau, longtime Eugene Higgins Professor of Physics and Natural Philosophy, Yale University, also thinks that our mind (soul) is not ravished by time:[37)]

> "If my considerations are correct, each individual mind is part of God or part of the Universal Mind. Much as our thought can survey and come to know all space, the Universal Mind can travel back and forth in time at will. Its universality comprises not only the mind of each of us but is equally aware of our past."
>
> Henry Margenau

Indeed, Plato thought time was a derived quantity; i.e., he maintained that before the big bang there was no time. In Plato's own words:

> "Now when all the stars which were necessary to the creation of time ... [and] the birth of time ..."
>
> Plato in *Timaeus*

Key Points

- $S = k \log W$.
- Real engines (steam, gas, etc.) are not reversible, and are inherently inefficient. Only an imaginary engine (which operates infinitely slowly) is reversible.
- Universal entropy always increases and thus provides an "arrow of time."

5
Maxwell's Demon and Szilárd's One-Atom Engine
A heady mixture of entropy, information, and consciousness

James Clerk Maxwell (1831–1879)

Leo Szilárd (1898–1964)

"Now let us suppose that such a vessel is divided into two portions, A and B, and that a being, who can see the individual molecules, opens and closes a hole, so as to allow only the swifter molecules to pass into A, and only the slower ones to pass into B. He will thus, without expenditure of work, raise the temperature of B and lower that of A, in contradiction to the second law of thermodynamics."

James Clerk Maxwell (1871)

"In his paper (*Zeitschrift für Physik* **53**, 840, 1929) Szilárd constructs a perpetuum mobile of second kind by trading in entropy with knowledge. This seemingly crazy idea of equating something material with something purely spiritual turned out to be very fruitful and led in the 1930s to Shannon's information theory."

Walther Thirring in *Leo Szilárd Centenary Volume*

The Demon and the Quantum, Second Edition. Robert J. Scully and Marlan O. Scully

ISBN 978-3-527-40983-9

5.1 Maxwell, the Man

The bright red wooden cone wound and weaved like a speeding yo-yo. The boy working the two strings it rolled on was entranced by the cone's magical, dancing motions. By twisting, pulling and crossing the two strings, he made the shiny cone glide, flow, even jump off the string and somersault. Clearly, he had plenty of experience doing this, so expertly did he control the butterfly-like motions of the wooden cone. The game was known as the "Devil on Two Sticks" and it was soon to spread in popularity around the world. The soft flowing noise it made mingled with the breeze blowing in from the picturesque highlands to the north. "James," called the older man from the porch of the nearby house, "It's dinner time." Mr. Maxwell had recently struggled with uncertainty following the death of his wife, and he worried about how little James would handle the loss of his mother. But a year after Mrs. Maxwell's death, James was well on his way toward becoming a calm and happy boy. Now nine, he had grown especially close to his father.

Intelligent and inquisitive, the young boy was always trying to find out how things worked. Not satisfied with vague explanations to his frequent questions of "What's the go o' that?," he would usually get his father to offer a more detailed explanation.

As fate has it with so many things in life, people sometimes get credit for either far more, or far less, than they deserve. A few will have extraordinary feats attributed to them which they did not truly achieve. But history has not been fair to James Maxwell (1831–1879).

Some have said that his laws of electromagnetism furthered science more than Newton's laws of motion and Einstein's theory of relativity combined. Yet few outside of science have heard of the name James Clerk Maxwell. With his cheerful disposition and strong, yet humble, religious faith, it is doubtful if he would have cared. The true greats are often like that, desiring nothing but the thrill of discovery as their reward. The contributions of Maxwell that we are most concerned with involve the theory of heat. His insights into thermodynamics were on a par with those of Sadi Carnot. Maxwell's research provided the final step necessary in the emerging field of statistical physics before the mathematical definition of entropy was coined by Boltzmann. In fact, it was Maxwell's goal from early on in his career to better understand entropy. In attempting to do this, he discovered the physical laws governing the thermal behavior of particles in gases and fluids.

How can it be that the single most important source of ideas and insights of our present book, a man who ranks with Newton and Einstein, seems under-appreciated in his home country? For example, as Basil Mahon notes: "When the Royal Society of London held its tricentenary celebration in 1960, the Queen praised a number of former Fellows in her speech. However, Maxwell was not on the hero's list!" Is it possible to understand such an outrage? We must try.

First, he was something of a rustic, a country bumpkin, his early life spent on the family estate in remote Galloway, Scotland. Arriving at the Edinburgh Academy in 1841, at the age of ten, his fellow students took pleasure in tormenting him over his back-country accent, homemade clothing, and shy demeanor. He was beaten by gangs and ridiculed. They called him "Dafty." Later he bested his tormenters in vigorous fist fights, but the nickname stuck.

Over the next decade, Maxwell did well at Edinburgh. He made important lifelong friends such as Lewis Campbell and P. G. Tait, both of whom became successful scholars. But he was always considered a bit odd. As the mother of his friend, Campbell, wrote in her diary:

> "His manners are very peculiar; but having good sense, sterling worth, and good humor, the intercourse with a college will rub off his oddities. I doubt not of his being a distinguished man."

In 1850, Maxwell went to Cambridge, the Mecca of English higher education. There, he did very well. He was a top student in mathematics and physics. He was even more distinguished as a young researcher. For example, he showed how all colors relate to the three primary colors by the Maxwell color triangle.

Unfortunately, his father became ill in the winter of 1854 and died two years later. Young Maxwell returned to Scotland to care for the family estate and its people. He obtained employment at Marischal College in Aberdeen as its youngest professor. There he did ground-breaking research into several topics, notably the kinetic theory of gases. However, it developed that the two colleges in Aberdeen, Marischal and King's College, fused into one. Only half the number of professors were needed. Incredible as it may seem, Maxwell was among those sacked.

Here we have an important hint into the Queen's error or oversight. When first hearing of the Queen's outrageous neglect of Maxwell, one is tempted to think that some sort of English–Scottish bias was at work. However, Maxwell did not overwhelm even the Scots. The professorship in the combined Marischal–King's College went to David Thompson. Though professorships were in short supply in Scotland, a new post opened up in Edinburgh University. Surely, Maxwell would get the nod for this post. But no, instead, Maxwell's old friend, Tait, beat him in the competition for the chair at Edinburgh.

Indeed, the English seemed to value him most. Shortly after Aberdeen dismissed him, he received a call to be a Chair in King's College, London. He accepted, and held this position until he resigned in 1865 to devote time to his estate and write his famous "Treatise on Electricity and Magnetism."

It is here, in the study of electricity and magnetism, that Maxwell really hit pay dirt. He showed from his unified equations for electricity and magnetism that electromagnetic waves exist. These waves travel at the incredible speed of 186 000 miles per second. Since this speed is the same as the speed of light, he concluded that light was also an electromagnetic wave. We call the speed of light c and it is useful to remember that it is approximately 3×10^8 meters per second. This is the c in Einstein's famous equation $E = mc^2$.

This, then, was Maxwell's biggest contribution. He showed that light consists of electric and magnetic fields propagating through space, each supporting and generating the other as they whiz along. This earth-shaking discovery is enough to earn Maxwell superhero status, in the same class as Newton and Einstein.

He made many trips to Cambridge during the time at his estate, for example, to serve on the Tripos Exam Committee. In 1871, five years after leaving London, he accepted a post at Cambridge as the first Cavendish Professor of Physics. He held the position until he died of cancer in 1879 at the age of 48.

He was a giant. So why does he not rate a mention on the Queen's scientific big hitter list? For one thing, he died too soon.

It was some ten years later that Heinrich Hertz proved he was right. Just as Eddington's deflection of light experiment excited the public about Einstein's theory of relativity, so Maxwell would have been a (bigger) hero if he had lived to see his theory verified by Hertz's famous experiment.

For another thing, and most importantly, he did not care. He was not in it for the glory, as he said at the end:

> "The only desire which I can have is like David to serve my own generation by the will of God and then fall asleep."

5.2 Maxwell Combines Intelligence and Thermodynamics

In the 19th century, scientists became baffled by their inability to explain the mixing of hot and cold fluids or gases. Of course, they could not explain this because they didn't understand or even thoroughly believe in the existence of atoms and molecules, the carriers of heat. The concepts of entropy, heat and energy are related; discovering more about any one of these phenomena provides a glimpse into the others. Maxwell was aware of these interconnections and he began his research by trying to understand the workings of heat and energy.

Maxwell's theory of heat helped to lay the foundation for the still emerging field of thermodynamics. For example, his curiosity was piqued by the cooling of a hot cup of water. Why did the heat always flow into cooler surroundings? Why can't cold temperatures flow into hotter ones? What laws prevent a cool cup of water from spontaneously warming up? These questions, perhaps not very profound on the surface, probe some of the deepest aspects of the subtle workings of nature.

Maxwell was intrigued by thermodynamics. He thought the random behavior of heat motion might be controlled by a conscious being, later called "Maxwell's demon." He described his benign demon as follows:[38]

> "[I]f we conceive a being whose faculties are so sharpened that he can follow every molecule in its course, such a being, whose attributes are still as essentially finite as our own, would be able to do what is at present impossible to us." Maxwell then notes that if a vessel full of gas

"is divided into two portions, A and B, by a division in which there is a small hole, and that a being, who can see the individual molecules, opens and closes this hole, so as to allow only the swifter molecules to pass from A to B, and only the slower ones to pass from B to A. He will thus, without expenditure of work, raise the temperature of B and lower that of A, in contradiction to the second law of thermodynamics."

As with many great scientific ideas, the ability of the demon to beat the second law was only hypothetical in its original form. But as a brilliant concept it served to guide us to the proper solutions to some very important issues. The existence of the demon has been disproved time and again in the past 100 years; however, the demon refuses to stay down.

Maxwell had sought to better understand the meaning of entropy operating at a microscopic level. The intelligent heat sorter was an interesting paradox but not a serious suggestion for a new heat engine. The idea of a demon was offered in that sense. Others pushed the idea further. For example, Marian von Smoluchowski[39] says that a device "appropriately operated by intelligent beings" might really beat the second law.

Notice that there is really something new here – the element of consciousness has, for the first time, entered into the realm of physical science. Maxwell's ideas of a controlling being may have been much further ahead of his time and more on target than anyone, including himself, ever suspected. The demon was, in some sense, a harbinger of things to come in our study of the microcosmic world. In the following we will look at a simple analysis of an engine operating by a measurement of atomic position, i.e., the Szilárd single-atom heat engine. As we will see, Szilárd's engine is an excellent introduction of the demon concept as well as a partial resolution of the problem Maxwell had posed.

5.3 Szilárd Sheds Light on the Demon Paradox by Equating Information With Negative Entropy

In order to study the role of intelligence in thermodynamics, Leo Szilárd (1898–1964) developed the idea of a single-atom engine which used *measurement* as a means of extracting useful work

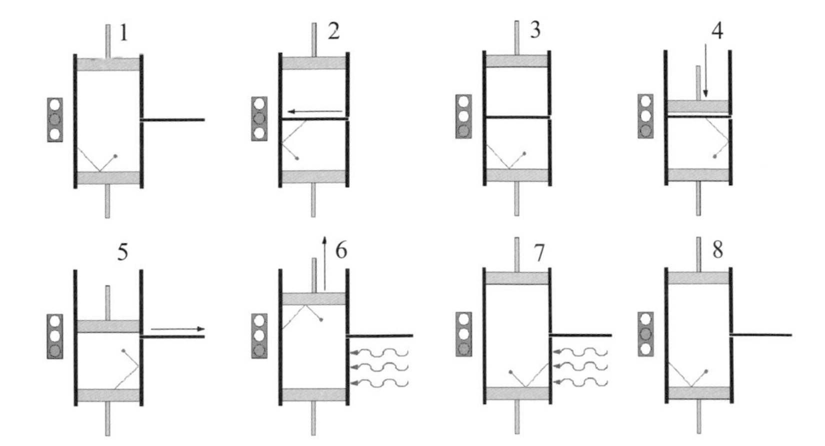

Fig. 5.1 In Szilárd's 1929 engine, (1) a single molecule is confined in a cylinder by two pistons. A movable partition can divide the cylinder into two compartments. At the beginning of the engine's cycle (2) the partition is inserted, trapping the molecule in one half of the cylinder. The observational device determines and records which half contains the molecule (3), and the piston from the other half is pushed in until it touches the partition (4). The partition is then withdrawn (5) and the molecule strikes the piston, pushing it up (6) as the one-molecule gas "expands" against the piston. Energy lost by the molecule as it works against the piston is replaced by heat from the hot energy source. When the piston has returned to its original position (7), the register is reset (8), and the cycle can begin again.

without a lower-temperature entropy sink (exhaust/compression stroke). This is much in the same spirit as Maxwell's sorting demon. In Fig. 5.1, the sliding partition in the middle of the cylinders is sorting the single atom, acting like a traffic-controlling demon now by detecting the atom.

In Szilárd's scheme (see Fig. 5.1) a single molecule is confined in a cylinder between two moving pistons. In this engine a movable partition divides the cylinder into two compartments. By using a monitoring device that measures the content of the two compartments, one may move the pistons in such a way that the engine seems to be converting the heat of the environment into work on *every* cycle.

The engine is simplicity itself. Although we don't know how to build it yet, it has much in common with other heat/steam engines. The hot atom represents the expanding gas that pushes the piston. The unique point of it all is the sliding partition and the measuring device, which would seem to cheat the second law of thermodynamics by eliminating the need for a lower-temperature reservoir acting as an entropy sink.

This interesting variation on the Maxwell demon theme threatens to provide a source of perpetual motion of the second kind. Szilárd tried to save the second law by arguing that it must be the act of measuring that requires energy and compensates for the entropy loss. He says:

> "One may reasonably assume that a measurement procedure is fundamentally associated with a certain definite average entropy production, and that this restores concordance with the Second Law."

In other words, he believed that measuring the position or location of the atom would generate entropy and require energy.

Szilárd was unable to prove this statement and it was later proven wrong. The act of measurement *can* be performed without doing work. The step that cannot be performed without doing work is the discarding, or erasing, of the information stored in the measuring apparatus. The erasure of information is analogous to the exhaust stroke in a steam engine. Erasing information takes energy. This is so because erasing is an irreversible process. In Landauer's words:[40)]

> "Computing machines inevitably involve devices which perform logical functions [generating heat], typically of the order of kT for each irreversible function."

Hence, at the end of each cycle one needs to reset the memory register. This involves a logically irreversible process. By resetting the register, one converts a state of two possible values (top, bottom) to a single neutral state. This erasure lowers the entropy of the register by $k \log 2$. This correspondingly increases the entropy of the engine's surroundings by the same amount. Hence, the waste work associated with resetting the register is $kT \log 2$. The useful work was $kT \log 2$ also, hence the net work gained in a cycle is zero. This result is in accord with the second law of thermodynamics.

The following statement from Charles Bennett in a 1987 *Scientific American* article makes the point well:[41)]

"We have, then, found the reason the demon cannot violate the second law: in order to observe any single molecule, it must first forget the results of previous observations. Forgetting results or discarding information is thermodynamically costly."

The differences between a single-atom Carnot engine and a single-atom Szilárd engine are summarized in Fig. 5.2 and Tab. 5.1.

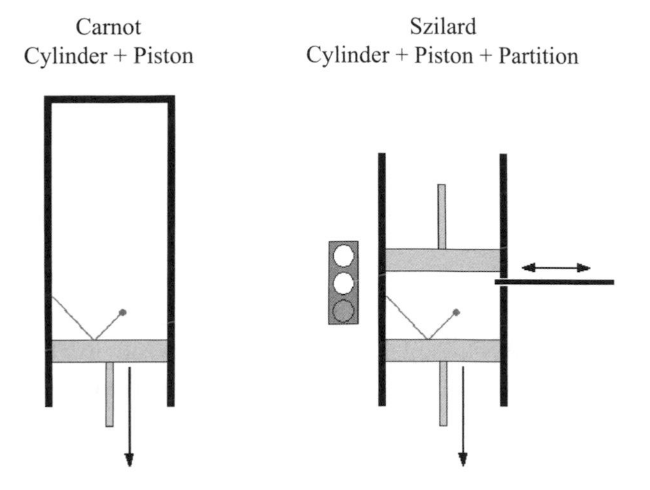

Fig. 5.2 Illustrations of the single-atom Carnot engine and the single-atom Szilárd engine.

Table 5.1 Comparison of single-atom Carnot and Szilárd engines.

	Carnot engine	**Szilárd engine**
Useful work	Power to shaft	Power to shaft
Working fluid	Atomic motion	Atomic motion
Energy source	Hot bath (e.g., nuclear reactor or coal furnace), T_h	Hot bath (boiler), T_h
Entropy sink mechanism	Cold bath (e.g., cooling tower or automobile radiator), T_c	Reset monitor
Entropy discarded via	Compression at T_c	Erasing information

Thus we see that Maxwell's intelligent demon led Szilárd to introduce the notion of information into science. At the same time, people like Einstein, Schrödinger and Wigner were led to consider the role of observation and intelligence in the micro-quantum world, as we will discuss later. In the next chapter we shall begin our journey into the quantum micro-world in order to better understand the Maxwell demon paradox.

Key Points

- The Szilárd engine must discard entropy of amount $\Delta S = k \log 2$ on each cycle.
- This means waste work of amount $T\Delta S = kT \log 2$ is required on each cycle.
- The amount of good (useful) work per cycle is $W_{\text{useful}} = kT \log 2$.
- The net useful work per cycle is thus zero and the second law is not violated.

6
Quantum Mechanics I

From micromagnets to micromasers

Julian Schwinger rides a hay-wagon at the Scullys' New Mexico ranch.

Charles Townes examines a rare woodpecker at the Texas ranch of George Mitchell. (Photo courtesy of Roland E. Allen.)

> "The hot system has got hotter and the cold colder and yet no work has been done; only intelligence of a very observant and neat-fingered being has been employed. If one could only deal with the molecules directly and individually in the manner of this supposed being, one could violate the second law. 'Only we can't,' added Maxwell, 'not being clever enough.'"
>
> Harvey Leff and Andrew Rex[42)]

The Demon and the Quantum, Second Edition. Robert J. Scully and Marlan O. Scully

ISBN 978-3-527-40983-9

6.1 Setting the Scene

As promised at the end of the last chapter, we now proceed to show how Maxwell's "intelligent neat-fingered being" can be realized, and the paradox resolved with quantum mechanics. To that end, we will develop another kind of single-atom engine. But unlike the case with Szilárd's single-atom engine, we will make use of quantum mechanics to produce a real demon-like, hot–cold sorter.

Such a device was actually built by Otto Stern and Walther Gerlach in the same time frame as Szilárd was thinking up his single-atom engine. Furthermore, the Stern–Gerlach (SG) device can be used to transform thermal radiation (hot energy source) into coherent maser/laser radiation (useful work). Moreover, we shall find that cyclic operation involves an erasure of potential information, which requires work.

First we must learn a little quantum mechanics. The ideas we will be exploring in Section 6.2 are the basis for a path into quantum mechanics that Julian Schwinger (who shared the Nobel Prize with Feynman and Tomonaga in 1965) used when he taught introductory quantum mechanics.

Then in Section 6.3 we will tell the story of how Charles Townes and his student Jim Gordon used a Stern–Gerlach type of apparatus in order to make a real (NH_4) maser. Norman Ramsey and his student Dan Kleppner also made a maser out of hydrogen atoms using a hot–cold beam sorter. Finally, we will explain how Herbert Walther and Serge Haroche have drawn upon similar ideas to make amazing, single-atom masers. Taken together, these various advances allow us to envision, in the following chapter, a new kind of single-atom engine, one that operates via erasure of potential information.

6.2 The Schwinger Approach to Quantum Mechanics Using Atomic Micromagnets

If you can't join em, beat em.

Julian Schwinger (1918–1994)

Once, playing baseball as a young child, Julian Schwinger listened to his older brother Harold coaching him from the sidelines. When he was up to bat, he heard him call "Swing!" amid much laughter. Julian did swing – and missed. The pitcher wound up again and winked at the older brother. As soon as the ball left his hand ... "Swing!" Julian swung and missed the second and then a third time. Walking over to his brother, Julian asked Harold, "Why did you tell me to swing at those bad pitches?" The older brother laughed, "You're on the opposite team; I wanted you to strike out!"

This was a simple childhood event, but Julian was far from simple. He decided from that moment on to think for himself – he would call his own shots and listen to no one. And he did continue to think for himself. Indeed, when discussing his childhood on a relaxed ski trip, he cited that incident as one that set his world view.

A child prodigy and independent thinker he was, but distant and cold he was not. For example, once when he came to visit our New Mexico ranch, his car got stuck in the mud on one of the many unpaved, poor quality roads leading to our house. I got behind his car and pushed, heaved and shoved while the car inched out. His wife, Clarice, cheered me on from the passenger seat and waved at me when the car finally got enough speed to move along on its own (when once you finally get moving in the slippery, wet clay-like mud found in many parts of the state, you don't dare stop, or you will find yourself stuck again). What would be a frustrating inconvenience to others was a lark to them. It takes a sunny disposition to get a kick out of being stuck in New Mexico mud.

The Schwingers were open, friendly people who valued and respected others. While at a European conference dinner for Nobel Prize winners, Clarice pointed out Heisenberg standing alone. He had been one of Germany's top scientists, spending the war heading up the German atomic bomb project. No one was inviting him to sit with them, as involvement with the Nazis was a mark of intellectual leprosy. But at the suggestion of Clarice, Julian invited him to their table. Human and kind, the Schwingers were always thoughtful of others.

So here's a glimpse into the mentality of a great quantum theorist, brilliant but considerate, while always remaining open-minded and observant. He was committed to science all his life. Early in life, he appreciated the deep subtlety that is quantum reality. Norman Ramsey recounts the occasion in which he and Nobel Prize winner I. I. Rabi were discussing a quantum problem and 16-year-old Julian made a useful remark from the back of the room. Rabi was impressed and said something like, "Who is this kid?" Rabi became Julian's mentor shortly after that. An inquisitive, optimistic mind is best prepared to make the most of every problem. Julian understood this and made a living resolving quantum puzzles.

Julian taught quantum mechanics from the ground up. Very little (in fact, no) previous knowledge of the subject was assumed in the introductory course he offered. His approach is perfect for our purposes. His genius for math and physics did not prevent him from reaching out to those without such extensive training. He put it this way:

> "I propose to present an ideal induction of the general laws of quantum mechanics from a well-selected set of experiments – indeed, from a single type of experiment [that of Stern and Gerlach]."
>
> Julian Schwinger

The key point here is that atoms often behave like tiny magnets, just like the doggie bar magnets we see in toy stores and physics lecture halls, but much smaller. Doggie bar magnets, once a popular children's toy, consist of two little plastic dogs, each glued to a small bar magnet. The magnets will repel each other in one position, but then attract with impressive force if placed in just the right positions. They provide insight into the dynamics of small magnets. For example, hydrogen and silver atoms have a tiny magnetic moment (Fig. 6.1), which scientists can, and do, use to push them around in order to learn more about the quantum microworld. This is just what Otto Stern and Walther Gerlach did; what they found was a big surprise. In this chapter we shall explain how the Stern–Gerlach experiment works and provide an example to demonstrate its operation.

The basis of the Stern–Gerlach experiment is the deflection of the atomic magnet just introduced. This is a microscopic imitation of the little bar magnet, mentioned earlier, that you may have played with as a child. You may recall that the force on one magnet produced by another depends on their relative orientations. I used to try and try to make the magnets go together in just such a way as to overcome relative orientation, but they were absolutely fool-

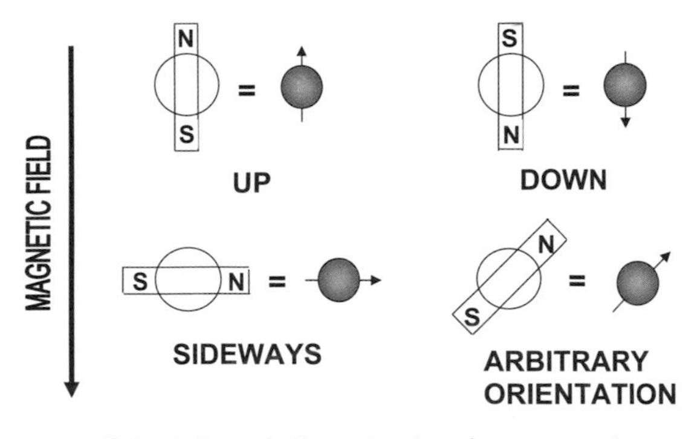

Fig. 6.1 Silver atoms have a magnetic moment, so we need to know the direction in which a magnet is pointing as well as how fast it is going, to totally specify its state of "being".

proof, deflecting and then coupling up in the only way their magnetic orientations would allow.

As indicated earlier, the experiment used by Stern and Gerlach consisted of shooting their tiny atomic dipole magnet (a silver atom) through a magnetic field, like that produced by the bar magnet of Fig. 6.2, and looking at the deflection of the atoms. But their experiment employed a better field design than a bar magnet. They

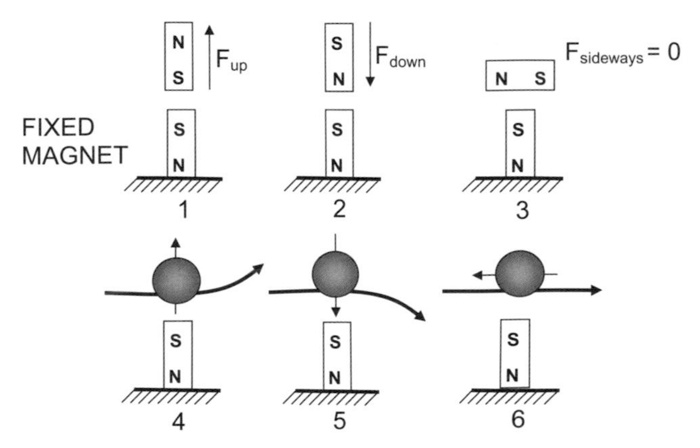

Fig. 6.2 Deflection of a floating bar magnet and a silver atom by a fixed bar magnet.

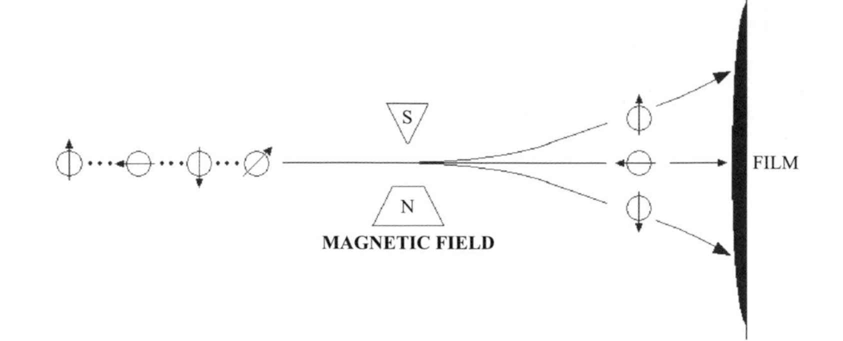

Expected deflection pattern of randomly oriented silver atoms

Fig. 6.3 This is the deflection pattern "expected" (on classical grounds) of randomly oriented silver atoms shot through a non-uniform field at a film target. Notice how they are deflected according to the orientation of their poles (north, south, in the middle – or if we showed more atoms, they would presumably be all over the film shown in the diagram). The amazing thing is, it doesn't work that way!

used a triangular and trapezoidal magnet as sketched in Fig. 6.3. They expected to see a uniform smear of atoms on the screen (usually a piece of film) as in Fig. 6.3, but that is not what they saw. Instead, they saw the surprising pattern seen in Fig. 6.4. It is as if the microscopic, silver atom magnets point only up (parallel to the direc-

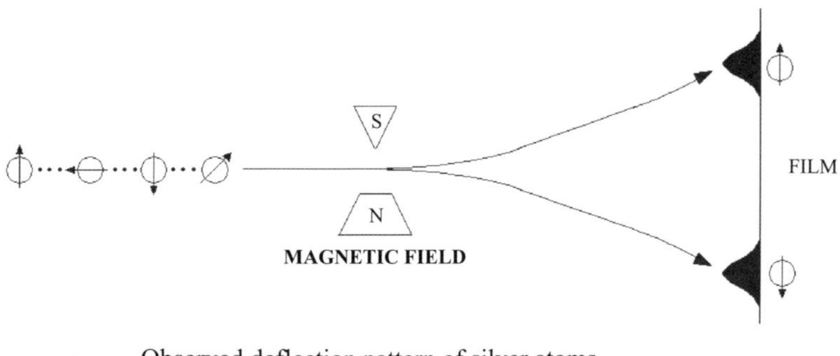

Observed deflection pattern of silver atoms

Fig. 6.4 This is the observed deflection pattern of randomly oriented silver atoms shot through a non-uniform field at a film target.

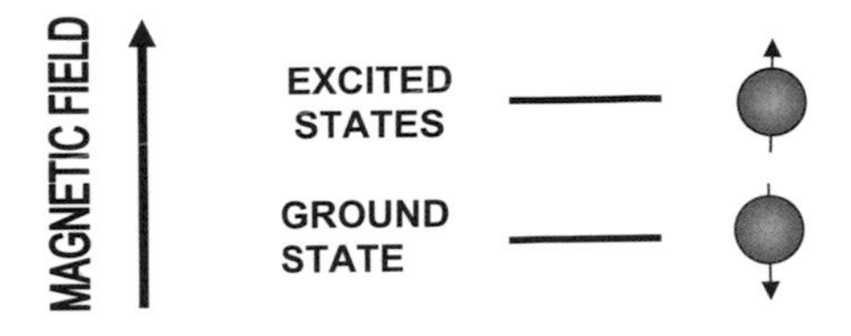

Fig. 6.5 The internal excited state is populated (back in the oven) by, for example, collisions with faster atoms having higher temperatures. We properly call excited atoms "hot" and their ground state "cold".

tion of the field) or down (anti-parallel to the direction of the field).

A uniform magnetic field will not affect the deflection of the silver atoms because it exerts the same force on both the north and south poles. The non-uniform field produced by the trapezoidal magnet does that. However, we assume that there is also a uniform field in the vertical direction. The atomic magnetic moments want to be aligned anti-parallel to the field since this is the state of lowest energy. The atoms in the parallel alignment, on the other hand, have higher energy. This is summarized in Fig. 6.5. The excited

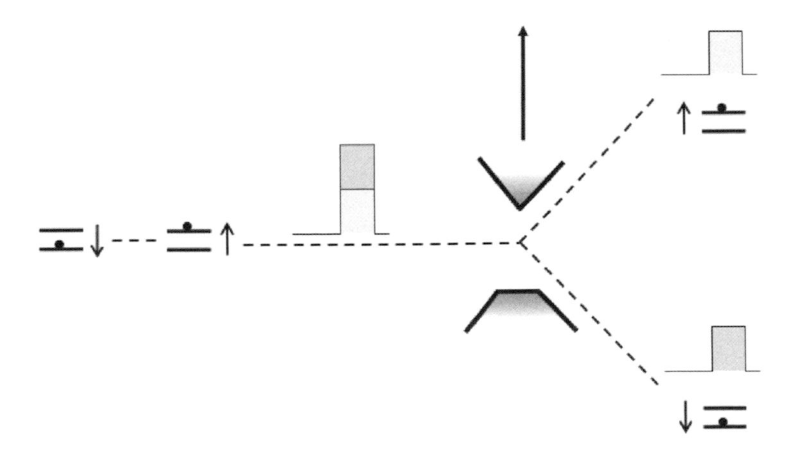

Fig. 6.6 Hot (spin-up) atoms are deflected one way and cold (spin-down) atoms are deflected the other way in a Stern–Gerlach apparatus.

(hot or spin-up) atoms are deflected one way and the ground-state (cold or spin-down) atoms the other way (Fig. 6.6). Maxwell would surely have been amused to see this counterpart of his "intelligent" hot–cold sorter.

6.3 How Charles Townes and Team Used a "Hot–Cold Sorter" to Generate Coherent Maser Radiation

The "hot–cold sorter" was used to make the first maser (which stands for microwave amplification by stimulated emission of radiation).

Charles Townes, like many other scientific greats, showed exceptional promise at a young age. When he was a boy, his Christmas list consisted of tools from the general store to help him with experiments he was doing at home. A highly motivated student, the Depression did not stop him from obtaining a degree. Born in 1915, he was destined to be among those of the great generation who strove and fought to lead America into some of its best years after World War II. Townes received a Ph.D. from the California Institute of Technology when he was 23. In 1939, he joined Bell Labs in New York City, where he contributed to the war effort with his research in radar.

Pearl Harbor had taught the value of radar. Part of the reason the Japanese caught America so off-guard was that we did not believe our own radar report of an approaching fleet of their fighter-bombers. The science of radar was in its infancy and not yet considered trustworthy, but after Pearl Harbor the best minds in the country were assigned to work on it. For example, William Hewlett (of Hewlett-Packard fame), future Nobel Prize winners and future science project managers of all sorts made their name in radar and microwave technology. America's success in World War II was aided by the developments in the new fields of microwave technology and electronics. After the war, the applications of microwaves led to many advances. Before the war, development was focused on basic science and a wide range of innovations made by young entrepreneurs. After the war, the face of progress changed; technology was progress itself and it was controlled by those who understood its subtle secrets. No longer would that progress occur in the basements of ambitious innovators. It was now tied to big research labs and the industrial/defense conglomerates spawned by the war.

These research laboratories included university laboratories. A prime example is the Townes group at Columbia where the laser/maser had its start. Townes' microwave/radar research during the war included development of a radar imaging device used toward the end of the war on B-25 bombers. Sought the world over for his expertise, Townes has been awarded professorships at Columbia, MIT and now at the University of California at Berkeley. He became vice-president and director of research for the Institute for Defense Analysis in 1959. He shared the Nobel Prize in 1964 with Alexander Prokhorov and Nikolai Basov of the Lebedev Institute in Moscow, for his invention of the maser.

The maser has progressed from a microwave device to a lightwave device, known as the laser (which stands for light amplification by stimulated emission of radiation). Today the laser is widely used in medical apparatus, computer technology, military hardware, and bar code scanning devices. Anything optical that scans, measures or records likely uses a laser, a testament to its quality as a "perfect" instrument of science. The use of laser light as a measuring tool often produces results as accurate and precise as physical laws will allow. And if the laser's status as a tool of precision and measurement does not qualify it as a perfect instrument of science, its numerous other benefits to mankind ought to insure that distinction. An example of the laser's use in medicine is laser retinal "welding", as well as a host of other applications.

The long list of honors accorded to Professor Townes by no means defines this unusual man. Not only did he win the Nobel Prize, but many Nobel Prizes have been based on his pioneering work involving lasers and masers (see Fig. 6.7). Witness also the 2005 Nobel Prize given to Roy Glauber, Jan Hall, and Ted Hänsch for contributions to quantum optics and laser physics. Indeed, the centennial poster celebrating one hundred years of physics in the United States features Charles Townes at the center of the poster.

What comes closer to defining this remarkable man is the family tradition of contributions to the spiritual and social well-being of all they came in contact with. For example, when "Charlie", as he is known to his friends, went to Sweden to collect his Nobel Prize,

Dudley Herschbach, Nobel prize in Chemistry (1986) "for their contributions concerning the dynamics of chemical elementary processes."

Willis Lamb, Nobel prize in Physics (1955) "for his discoveries concerning the fine structure of the hydrogen spectrum."

Norman Ramsey, Nobel prize in Physics (1989) "for the invention of the separated oscillatory fields method and its use in the hydrogen maser and other atomic clocks."

Charles Townes, Nobel prize in Physics (1964) "for fundamental work in the field of quantum electronics, which has led to the construction of oscillators and amplifiers based on the maser-laser principle."

Fig. 6.7 Nobel prizes based on (Stern-Gerlach like) atomic and molecular beam experiments include D. Herschbach, W. Lamb, N. Ramsey, and C. Townes.

he met Martin Luther King, Jr., who was awarded the Nobel Peace Prize the same year. On that occasion, King asked him, "Are you the nephew of Clara [Hard] Rutledge? She helped me so much in the early years." To which Charlie [Hard] Townes happily replied that she was indeed his aunt. It is precisely this concern for others that characterizes our hero.

As mentioned earlier, the maser was the predecessor to the laser. The maser was and is really not so different in theory from a radio oscillator. Radio antennas generate radio waves by sending a current up and down the antenna, governed by a driving voltage. The oscillating electrons radiate radio waves. Masers create *microwaves* (centimeter length waves) while lasers create *light* (micrometer length waves). Inside the laser and maser devices are not electrical windings, wires and an antenna, but a cavity into which photons are radiated by atoms. The optical component of a laser consists of nothing more than a couple of mirrors. The atoms or molecules do the hard work (useful work); they radiate the light.

The technological changes set in motion by Townes and his colleagues are the foundation for much of today's science. The scientific laws and theories developed by Einstein and Maxwell were the foundation on which Townes built. For example, one might well ask how the maser stands with the second law of thermodynamics. After all, the maser seems to be generating useful work from only one heat source. We will have much more to say about this, but we note that Townes understood that the new device did not violate the second law, since it was not in equilibrium.

The stimulated emission which takes place in a maser is a process that occurs when a molecule or atom is in an excited state (like the "up"-pointing hot atoms of Fig. 6.6) and is then driven to the ground state by microwave (maser) or optical (laser) radiation. This is nothing more than stimulated absorption in reverse.

To better understand the correspondence between stimulated absorption and stimulated emission of photons by atoms, note that nature at the atomic level is essentially reversible. That is, if we take a movie of two atoms colliding and run it backwards, we still have a possible event (see Fig. 6.8).

This time reversal symmetry, as it is called, is true for the microworld but not the macro-world. The reason for the difference can be seen to involve entropy. Entropy is always increasing. In real

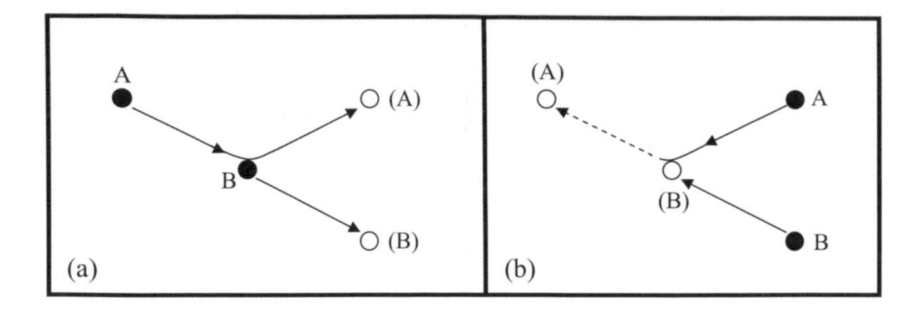

Fig. 6.8 (a) Atoms A and B before and after a collision. (b) The event after time reversal (movie running backward).

life, birds hatch out of eggs. We never see the backward, movie-like scenario of a baby bird returning to the egg in real life. The point is that, yes, time only moves forward. We all know that, but why does it only move forward? The reason is that entropy is always increasing, as required by the second law of thermodynamics.

Back to stimulated emission. Think now of an atom, which begins in the ground (cold) state and is "hit" by a photon beam. It will be excited, and absorption of one of the incident photons will take place, as in Fig. 6.9(a). Now let us time-reverse the stimulated absorption movie of Fig. 6.9(a). This will take the excited hot atom of Fig. 6.9(b) to the ground state and an extra photon will be emitted. This is stimulated emission.

Finally we note, following Townes, that if we inject hot excited atoms or molecules into a metal box (i.e., a microwave cavity) that holds photons, we can increase the number of photons in the cavity by stimulated emission. This is how the maser works (see Fig. 6.10).

Some 30 years after the maser was invented, researchers in Paris (led by Serge Haroche) and in Munich (led by Herbert Walther) developed maser cavities which were so well engineered they could hold microwave photons for a long time. That is, the photons in these new (superconducting) cavities could live for up to a second – a long life for a photon, which is usually lost from most cavities after some millionths of a second.

It then became possible to make a single-atom maser, that is, a maser which operates with only one atom at a time in the cavity. These amazing masers are now called micromasers. How Szilárd

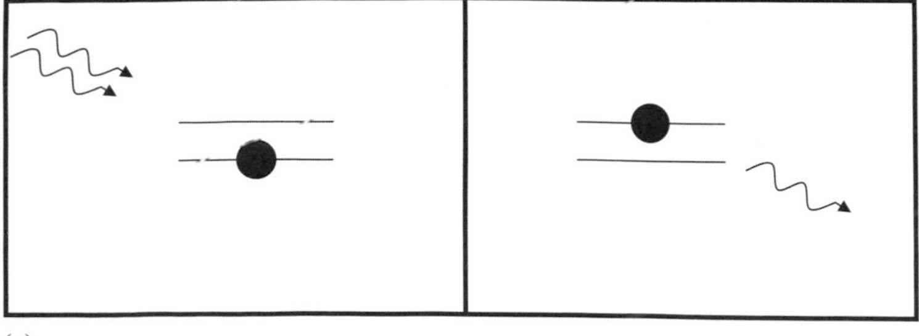

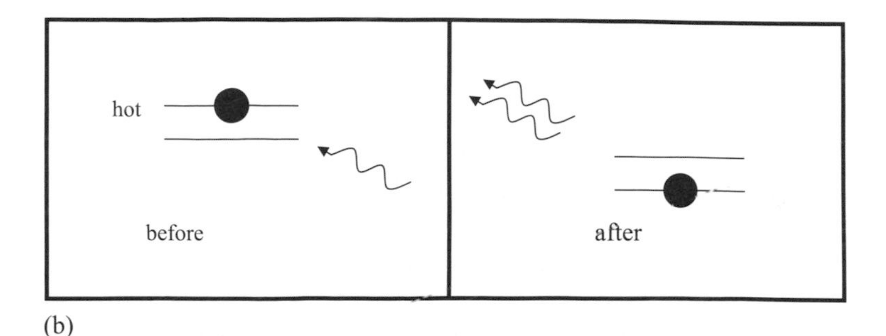

Fig. 6.9 (a) Stimulated absorption takes place when a cold (ground-state) atom is hit by incident photons, and absorbs a photon. (b) Time-reversed movie of the stimulated absorption in (a) yields stimulated emission. That is, an excited atom emits a photon by stimulated emission and goes to the ground state. Now the roles of "before" and "after" are reversed.

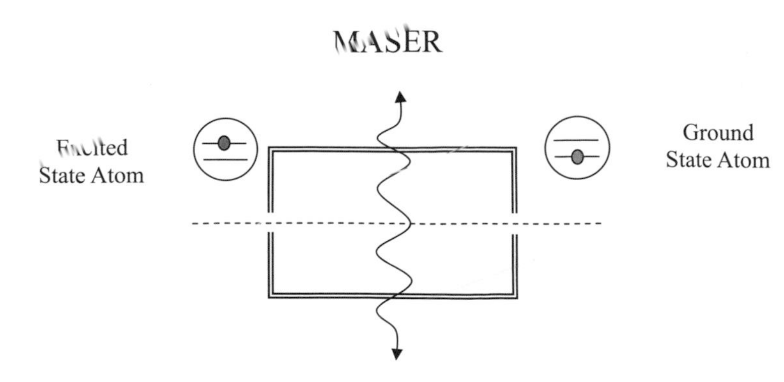

Fig. 6.10 The internal atomic energy of a hot atom is transferred to the field as the hot atom passes through the cavity and is stimulated to emit light or microwaves.

would have loved these single-atom micromasers. Furthermore (as is discussed at length in Endnotes 6), the maser typically operates with a mixture of hot and cold atoms at temperatures T_{hot} and T_{cold}. Instead of passing a beam of atoms through a cavity, physicists often put a batch of atoms in the cavity and bathe them in "hot" and "cold" light. Guess what? The maser now operates as a (quantum) heat engine with Carnot efficiency! More on this will be found in the next chapter and in Endnotes 6. Next, we will make a Maxwell's demon, single-atom engine by combining the Stern and Gerlach atom sorter with the single-atom maser.

Before proceeding to this topic, we recap the significance of the technology previously described. This chapter has given a brief and basic look at the workings of quantum physics. The peculiar deflection patterns shown in Figs. 6.3 and 6.4 demonstrate one of the mainstays of quantum physics. The facts of life depicted in those figures represent a radical departure from classical physics.

Today, modern quantum physics is aided by supercolliders, electron microscopes, and high-power lasers, a reminder of the guidance set forward by Brahe, whose work made it clear that new discoveries would be facilitated by carefully applying and improving on available technology. An exception that proves the rule is the work of Max Planck (1858–1947), who needed no new technology to found quantum physics. Planck had been told that there was nothing new to be discovered in physics. Mechanics and thermodynamics were finished (it was thought). Electricity was not thoroughly understood, but close enough (or so many thought).

In the early part of Planck's life, it was still not popular to believe in atoms. Early on in his career, Planck himself did not, an irony for the one who would eventually introduce modern physics. For atoms are to quantum physics what numbers are to mathematics.

Planck was clearly on the right track when he developed the formula for the entropy of thermal radiation. From the start of Planck's career, Clausius' idea on entropy had been central to his research. Thermodynamics and heat theory were one of Planck's early passions. He expanded this interest into the problem of black-body radiation. A black body is a perfect absorber, i.e., it looks black. Such a cavity is also a perfect radiator. Planck was trying to determine how the temperature of a body governs the intensity of electromagnetic radiation emitted by a black body. One goal was to come up with a better design for light bulbs for the German electric companies. His interests in entropy and his research in black-body radiation helped him to form a revolutionary idea about nature. He developed a theory that energy did not come in arbitrary amounts. Instead, he found that it came in discrete packets called quanta.

This is where quantum physics began. Planck's work led Einstein to discover that electromagnetic energy, E, was a multiple of a unit, described as $E = hv$, where h is Planck's constant and v is the frequency of the radiation. This cannot be reconciled with classical physics: Planck's equation, with h known as the famous Planck's constant, was the dawn of quantum physics. When the subtle nature of quantum physics begins to baffle us, consider this quote from a frustrated Max Planck, who spoke of his inability to grasp the meaning of energy quanta:

> "My unavailing attempts to somehow integrate the action quantum into classical theory extended over several years and caused me much trouble."

At least he was honest! Planck was not overly concerned with personal recognition or job security, nor was he involved with politics. Further proof of his integrity is provided by his statement about the utility of young scientists when he himself was older. He said:

> "A new scientific truth does not establish itself by its enemies being convinced and expressing their change of opinion, but rather by its enemies gradually dying out and the younger generation being taught the truth from the beginning."

Predictably, a man as open as Planck made few enemies, a rare quality among German intellectuals during those turbulent years between and during the world wars.

The virtues of this honest and dedicated worker never failed to earn the admiration of his friends. He was a friend of Einstein, playing music with him. Although he retired from Berlin University in 1926 (succeeded by Erwin Schrödinger), Planck stayed active in physics. He lectured extensively and became President of the prestigious Kaiser-Wilhelm-Gesellschaft (KWG). After World War II, the name was changed to the Max-Planck-Gesellschaft (MPG).

During the period of Hitler's crimes, he was a voice of reason, trying to avoid conflict with the Nazi regime while hoping things would improve for his Jewish colleagues. However, with the war's end, Planck would find himself categorized by Einstein and a few others as being one of many Germans having unreasoned patience with the Nazis. But Planck was no Nazi. While he deliberately tried to avoid the wrath of the Nazis, he did appeal directly to Hitler in the case of the Jewish scientist Fritz Haber. Haber was the head of German chemical warfare in World War I. Hitler was not swayed by Planck's request; in fact Planck himself came under scrutiny. First, the Nazis put pressure on him not to seek a second term as President of the KWG. Next, he encountered hostility due to his acceptance of Einstein's theories. These theories ran contrary to Johannes Stark's Deutsche Physiks, also called Aryan Physics.

Planck was obviously a powerhouse in the German scientific community to have avoided serious Nazi persecution. For in 1945, his son Erwin was executed for taking part in the plot to execute Hitler. Hitler was known for brutal reprisals against family members for much less, but none were taken against Planck. He was particularly close to Erwin, but tragedy was not new to Planck. His son Karl was killed in action during World War I, while Erwin was a prisoner-of-war in French captivity. His other two children, twins Emma and Grete, both died the first time they gave birth. His first wife also died, possibly from tuberculosis.

In spite of these circumstances, Planck never seemed bitter. He moved on after his losses; for example, he remarried two years after the death of his first wife. He was an active scientist until the end, wisely refraining from politics and nationalism. In his opinion, the tragedy of the Nazi dictatorship was due to "the ascent of the rule of the crowds." As previously mentioned, Planck made few enemies. He was astute enough to keep the Nazis at arm's length. At war's end, he was sought by both East and West for his scientific knowledge and leadership. Unfortunately, the end of the war found Planck squarely in the Russian sector of Eastern Germany.

The Americans and English were aware of Planck's plight. There was no way the Russians were going to allow him to move to the West. However, there were no walls in place yet; the Iron Curtain had yet to fall. So late one night a group of Americans simply drove to Planck's home in Eastern Germany and took him, with nothing but the "clothes on his back." They drove him to Göttingen, where he had friends and family, but no coat, boots, or even a change of clothes. However, some of the students there took the matter in hand. Together they collected a box containing a warm "great coat," overboots, gloves, etc., as well as several shirts and pants, and took their thoughtful gift to Max. As one of those former students, Nobel laureate Wolfgang Paul* tells it,[43)] Planck was touched. He wept and said that this was the best gift he had ever been given. Well he might have felt that way, because in those difficult times, assembling such a care package was no mean feat.

A few years later, these same students were asked to serve as pallbearers at Planck's funeral. One might have thought that great laureates and government officials would be accorded the honor, but no. Planck's wish was that his special benefactors, his intellectual heirs, were to be given that special recognition.

Wolfgang Paul went on to relate the students' reaction. Of course, they were honored, and they took the whole matter very seriously; but as Bohr said, some things are too serious to be taken seriously. In that vein, Paul tells that the students alleviated the stress by joking that the smaller ones should be put at Planck's feet while the stronger ones should be put at the head where the weight would be.

Key Points

- The Stern–Gerlach magnetic sorter pushes hot (excited) atoms one way and cold (ground-state) ones another.

* Not to be confused with Wolfgang Pauli. In fact Paul tells the story that when he met Wolfgang Pauli; Paul said: I always wanted to meet my real part. He was refering, of course, to the fact that $i = \sqrt{-1}$ (see section 1.2) is an imaginary number.

- Townes and Gordon used such a sorter to prepare a beam of inverted atoms which they sent through a microwave cavity to make a maser.
- As we show in Endnotes 6, a maser is a new kind of engine, a quantum heat engine, which operates with Carnot efficiency!
- Following Walther and Haroche we learn that it is even possible to make a one-atom maser, i.e., a one-atom quantum heat engine.

7
Using Quantum Mechanics to Resolve the Maxwell Demon Paradox

Solving the demon problem with quantum mechanics but without information science, sweat or tears

Otto Stern

Walther Gerlach

Plaque Commemorating Stern-Gerlach Experiment (Photo courtesy of Horst Schmidt-Böcking.)

> "In February 1922, in this building of the Frankfurt Physical Society, Otto Stern and Walther Gerlach discovered the quantization of the magnetic moment of the atom. Many physical and technical developments of the 20th century rely on the important Stern–Gerlach experiment including nuclear magnetic resonance, atomic clocks, and lasers. Otto Stern was awarded the Nobel Prize in 1943 for this discovery."
>
> English translation of inscription on Stern–Gerlach memorial

The Demon and the Quantum, Second Edition. Robert J. Scully and Marlan O. Scully

ISBN 978-3-527-40983-9

7.1 Introduction

In Chapter 5 we told the story of how Szilárd, and later Bennett, addressed the demon paradox by using a combination of thermodynamics and information science. In particular, Bennett argued convincingly that it is the process of resetting or erasing the measurement register that generates entropy and preserves the second law. We also note with pleasure the excellent new book by Seth Lloyd on the same subject, which is mentioned again at the end of the chapter (see also Endnotes 7). I wish I had had it when I started this project!

Now, fortified by the Stern–Gerlach demonesque heat-sorter of the last chapter, we can design a new kind of single-atom engine that works by sorting hot and cold states of atoms. This is in contrast with the Szilárd single-atom engine, which used information instead of a demon hot–cold sorter to produce energy.

We will find that we never need to introduce the ideas of information and information erasure. We will ultimately come down on the side of Bennett and Landauer, as their point of view has some conceptual advantages. However, not everyone agrees. For example, the eminent statistical physicist Dirk ter Haar did not like "loose talk" about information entropy, and accordingly has declared:[44)]

> "[T]he entropy introduced in information theory is not a thermodynamical quantity and that the use of the same term is rather misleading. It was probably introduced because of a rather loose use of the term 'information.'"

One does not have to agree with his every premise to see that he has a point.

In the years following, others have disagreed with the information erasure resolution of the demon problem. For example, scientist W. Porod says:

> "There appears to be no physical reason to stress the operation of erasing a single bit or of throwing away logical information."

He and coworkers argue that "noise" is the key concept in resolving the demon problem.

So it would seem that we have multiple motives driving us, as well as multiple questions to answer. These include the following:

(a) Can we provide a resolution of the demon paradox without appealing to information and information entropy?
(b) Can we think of a way to "build" a simple, perhaps quantum mechanical, engine designed around a real demonesque hot–cold sorter?
(c) Is there a connection between the "intelligent being" of Maxwell and the external observer problem in quantum mechanics?

We will address these questions in detail and find that the answer to each is a resounding "Yes!"

7.2 Converting Energy From a Single Heat Bath (Incoherent Thermal Photons) to Useful Work (Coherent Maser Photons)

As the first step toward building our single-atom quantum engine, suppose that, as in Fig. 7.1, a single cold atom is injected into a

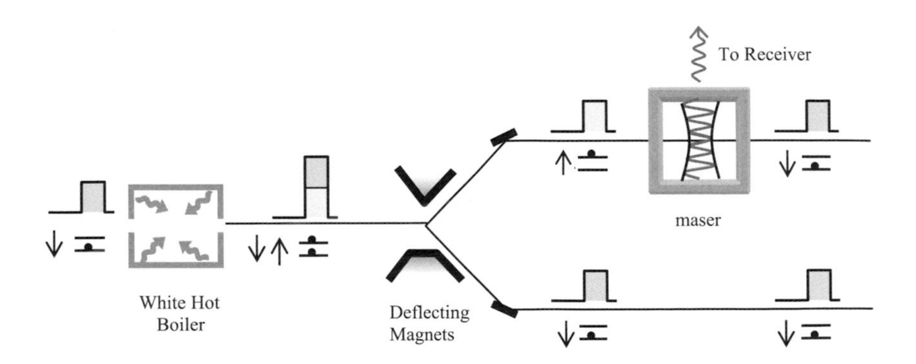

Fig. 7.1 A single source of heat energy and a Maxwell's demon sorter can produce coherent useful work (i.e., maser radiation). Note that this is accomplished using a single thermal bath, i.e., without using an entropy sink (lower-temperature heat bath).

white-hot cavity (boiler). The atom will be excited by the thermal photons in the cavity, and, after a short time, it will be in a mixture of the ground state and excited state. In particular, the upper and lower ground states will be equally populated for a sufficiently hot cavity. Let us assume that this is the case. Thus, we have roughly half the atoms coming from the cavity excited and half in the ground state.

Then the atom in Fig. 7.1 will be deflected up or down by the Stern–Gerlach demon with 50% probability. The hot atoms then pass into the maser cavity and are stimulated to emit coherent photons, i.e., do useful work.

So where are we? Are we violating the second law of thermodynamics? It might appear so, since we are using only a hot (energy) source but no colder (entropy) sink. However, we must recall that for a system to be a valid thermodynamic system it must operate in a cyclic fashion. Reconsider the steam engine. The steam (i.e., the working fluid) must cycle through the engine many times, going from the hot boiler, driving the piston, passing through the cold condenser, returning to the boiler, and then driving the piston again.

Thus, any engine must operate cyclically to be of any value. An engine whose pistons made only one power stroke would not sell very well. Yet that is what our demon-driven maser heat engine is at this point. This will not do. Suppose the Szilárd single-atom engine operated on only one cycle. Nobody would be very interested. We need to recycle our atoms by passing them back through the maser many times. Such a single-heat-bath engine operating by recycling our atoms is depicted in Fig. 7.2.

Thus the Stern–Gerlach demon/laser engine only works in a complete cycle when the upper and lower beams are *recombined* after each passage through the maser cavity and recycled through the "boiler" and then again passed through the maser. In this way we operate in complete thermodynamic cycles over and over. The connection with the second law of thermodynamics is not clear, but we have eliminated an exhaust (compression) cycle by a recombining scheme! How much (if any) energy must be spent in recombining the beams? To answer this, we ask how such a beam combiner might be built. This is the subject of the next section.

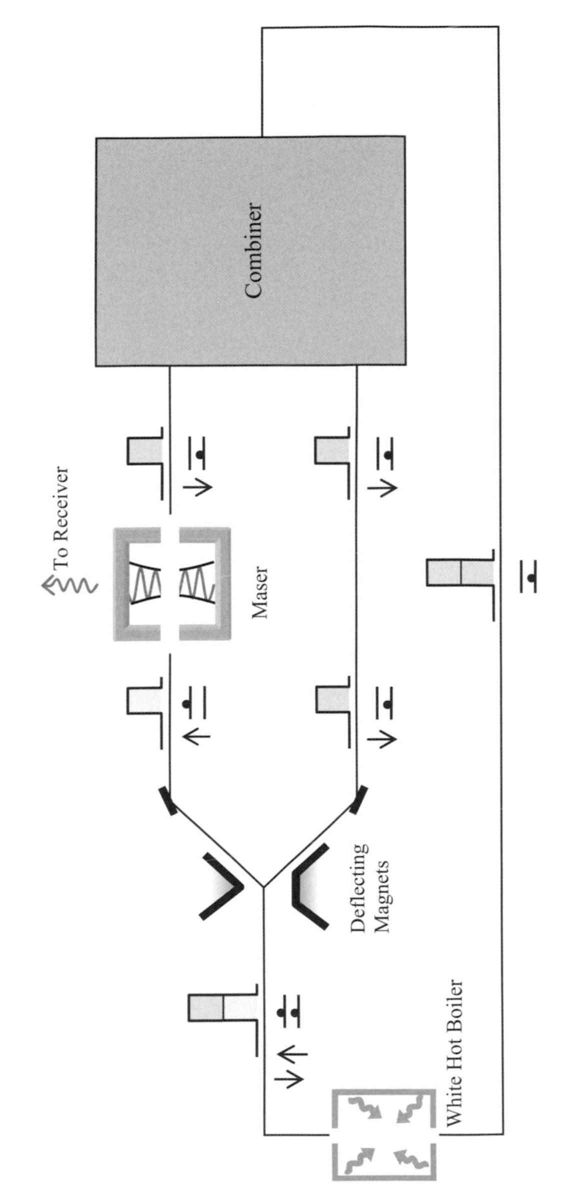

Fig. 7.2 A Stern–Gerlach (demon) maser functions by deflecting excited (hot) atoms into the upper path, thereupon passing through the maser cavity. The beam combiner allows atoms to be used over and over for many thermodynamic cycles.

7.3 How (and How Not) to Build a Beam Combiner

One might think that devising a beam combiner would be a simple matter. For example, we might think the "atom-in-a-pipe" model shown in Fig. 7.3 would do the trick. But, as we discuss below, the beam combiner of Fig. 7.3 does not work. In fact, this is an "old" conundrum. Electrical engineers and optical scientists learned it wouldn't work for photons in optical fibers. This is depicted on the cover of the book and explained at length in Endnotes 7. Scientists and engineers wrote papers on the subject with titles like "A fiber optical illusion," showing that you cannot stuff photons as they travel from two fibers into one (see Endnotes 7).

It is not hard to see why the combiner scheme of Fig. 7.3 will not allow our engine to operate in a continuous cycle. The clincher comes when we consider what the atom's trajectory would be after combining the upper and lower paths.

It might seem a simple matter to direct the atom as it goes along the upper or lower path into a Y junction and thus recombine the atomic trajectories, as per Fig. 7.4.

The problem is that the atomic trajectories associated with zig-zag (off the wall) paths due to the Y junction of Fig. 7.3 get us

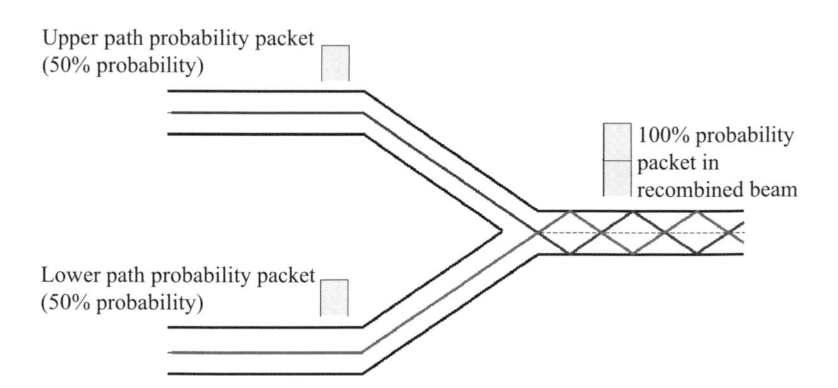

Fig. 7.3 The atom, depicted by the probability packet, for propagation along the upper and lower paths, and injected into the combiner pipe on the right.

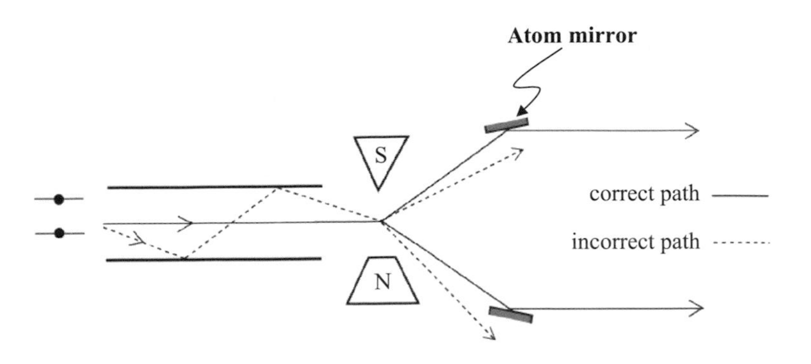

Fig. 7.4 The atom bouncing off the walls after passing through the Y junction of Fig. 7.3. The atom runs into trouble at the Stern–Gerlach demon sorter, leading to "incorrect" paths, as shown in the figure, and contrasted with correct paths.

into an unacceptable situation. That is, when we follow the atoms from the recombiner back to the Stern–Gerlach apparatus in order to go a second round, we are in trouble, as per Fig. 7.4. To put it another way, after the first pass through the maser, the atoms must go back through the hot boiler, get re-excited, and then pass into our Maxwell's demon sorter again. Then they must (again) be deflected into the maser cavity if they are hot and be deflected on the lower trajectory if they are cold. However, if the atoms are following a zigzag path as shown in Fig. 7.4, then they will not be deflected in such a way as to hit the (atom) mirrors at the desired angle. This means that the beam combiner of the type shown in Fig. 7.3 simply will not work.

However, it is possible to envision a beam combiner that does work. This is shown in Fig. 7.5. There, we see a rotating mirror which has the ability to transmit atoms moving along the top path and reflect atoms moving along the bottom path. The atoms moving along the bottom path are delayed so that they arrive a little later and see a reflecting segment of our rotating mirror as depicted in Fig. 7.5.

In Fig. 7.5 we have probability packets with the upper path (1) leading the lower path (2). They are then sent into the isothermal compression stage, as per the inset in Fig. 7.5. This arrangement

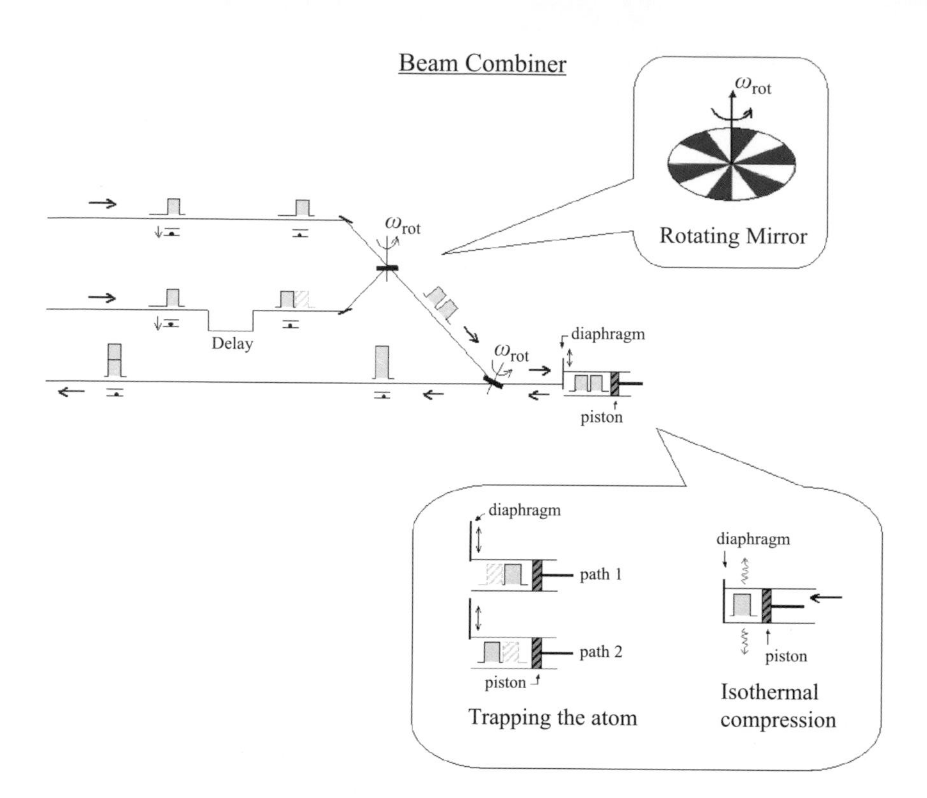

Fig. 7.5 The two trajectories (upper path and lower path) are recombined by using a rotating mirror which has alternating and equally spaced reflecting and transmitting sectors (upper inset). The rotating mirror is timed so that an atom in the upper trajectory is transmitted. However, when the atom is in the lower trajectory it is reflected. In order to accomplish this, a time delay is introduced in the lower path. After recombining, the probability packet of the atom (in the recombined beam) is spread out to twice its initial length. The probability packet is then isothermally compressed. In the lower inset we show the probability packet entering the compressor via upper and lower paths and its isothermal compression at the temperature T. The work required for the compression is actually greater than the useful work produced per cycle. See Endnotes 7 for more details.

enables us to trap the atom and to do isothermal compression (waste work) in order to push the double (twice as long) probability packet into the original configuration as in Fig. 7.5. Note that it has taken work to accomplish this compression or revitalization of our working atomic beam.

7.4 Pulling It All Together

Finally, we put the beam combiner of Fig. 7.5 into the Stern–Gerlach demon engine of Fig. 7.2 and we have the complete single-atom quantum engine of Fig. 7.6.

The energy balance can now be carried out. For any given pass, our atom will be in the excited state half the time, and since we leave energy in the maser field, of roughly kT (because the energy of the photon is approximately kT), we will leave $\frac{1}{2}kT$ average energy in the cavity on each cycle.

However, it costs work to compress the beam. As discussed in Chapter 4 and in Endnotes 7, the isothermal compression work (at temperature T) required is $kT \log 2$; since $\log 2$ is about 0.7, the waste work ($0.7kT$) is greater than the useful work ($0.5kT$).

So we are happy. We do use a demon heat sorter, but do not violate the second law with our demonesque engine. The reason is that work must be done to recombine the beams and *erase* the "which-way" (upper or lower path) potential information. Note that we never really "had" the upper/lower information. What is erased is the potential for knowing (or finding out) which path if we decided to look. The demonic (hot–cold) sorter is an essential part of our scheme, but the whole engine is benignly obedient to the second law of thermodynamics.

7.5 Conclusion

We have shown that a Stern–Gerlach apparatus can act as a Maxwell's demon by sorting atoms according to their internal states, hot atoms along one path, cold atoms along another. A heat engine that produces work by utilizing a Stern–Gerlach demon sorter can operate with a single-temperature reservoir. However, because it must operate in a closed cycle. This requires waste work in the form of compression of the atomic packet, i.e., erasure of "which-path" potential information. A comparison with the previous two single-atom heat engines of Carnot and Szilárd is given in Fig. 7.7 and Tab. 7.1. Clearly the present single-atom engine is very close to what Maxwell had in mind.

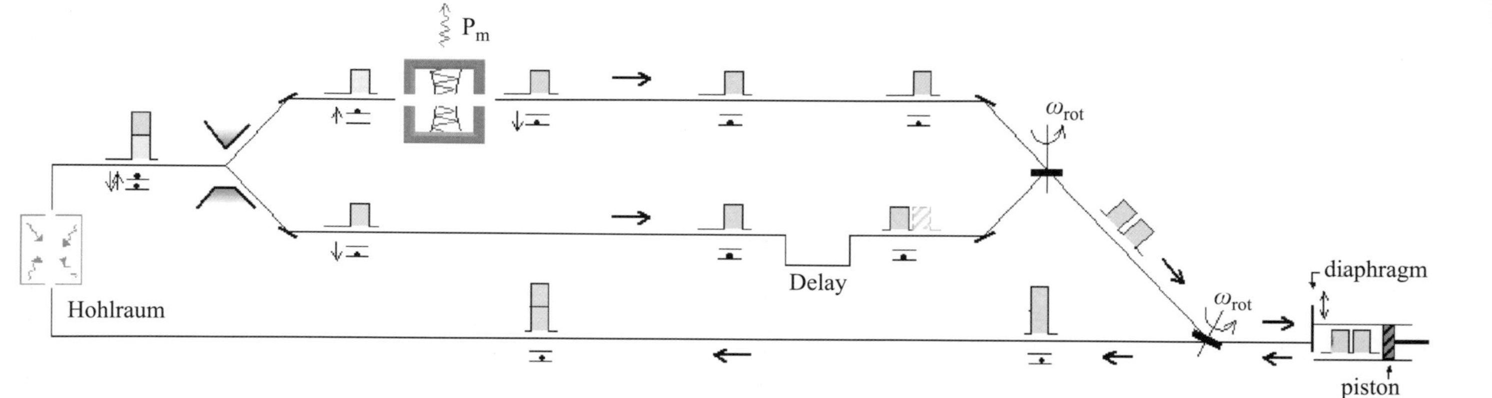

Fig. 7.6 A single two-level atom is prepared in a thermal mixture of its states by passing through a hot cavity containing thermal photons (i.e., a hohlraum). It is then directed through a Stern–Gerlach apparatus (SGA). The SGA acts as a quantum sorter. Hot (excited) atoms are deflected along the upper path and cold (ground-state) atoms along the lower path. If the atom is in the excited state, it passes through a resonant micromaser cavity where it emits a photon. This radiation is coherent and is therefore the *useful work* output of the engine (see Endnotes 7). On the other hand, if the atom is in the ground state, it follows the lower trajectory. There it undergoes a time delay. The two trajectories are recombined by using the alternating, sectored, rotating mirror as explained in Fig. 7.5. Finally, the atom (ground state) is passed through the hohlraum and thermally heated in order to complete the cycle.

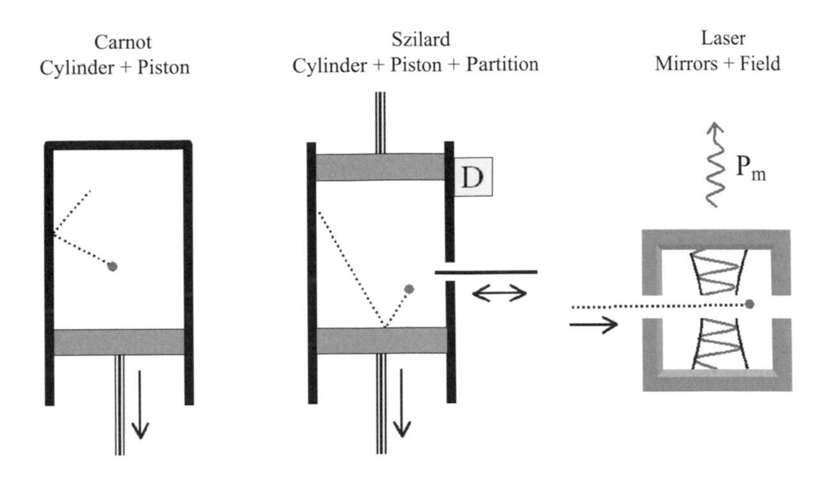

Fig. 7.7 Illustrations of the single-atom Carnot and Szilárd heat engines, and the single-atom laser quantum engine.

The similarities and differences observable with our toy single-atom engine compared with the Maxwell's demon problem are intriguing. The present single-atom engine has much in common with the Szilárd single-atom engine. Our preparation energy is in basic agreement with the Szilárd–Bennett result obtained on the basis of the theory of computing. However, no measurement is made in the operation of our single-atom engine. The present anal-

Table 7.1 Comparison of single-atom Carnot, Szilárd, and laser engines.

Engine	**Carnot**	**Szilárd**	**Laser**
Useful work	Power to shaft	Power to shaft	Radiation to receiver
Working fluid	Atomic motion	Atomic motion	Atomic internal states
Energy source	Hot bath (boiler) T_h	Hot bath (boiler) T_h	Hot bath (white hot boiler) T_h
Entropy source	Expansion at T_h	Expansion at T_h	Expansion of atomic packet
Entropy sink	Compression at T_c	Erasing actual information	Erasing potential information

ysis does not talk about information as such. Our engine does, however, operate by erasure of potential which-path information.

Thus our single-atom (quantum) engine has much in common with the single-atom Carnot and Szilárd engines, and resolves the Maxwell demon paradox using quantum mechanics. However, there was an earlier work, due to Seth Lloyd, exorcising the demon using quantum mechanics (see Endnotes 7). Unfortunately, Lloyd's *Physical Review* paper is beyond the scope of this book. Fortunately, as I mentioned earlier, he has just written a wonderful new book entitled *Programming the Universe* which explains his ideas. This book is on my highly recommended list.

Finally, we note that the concept of "erasure" was introduced by Scully and Drühl in quantum studies at about the same time Landauer and Bennett were considering the erasure of information in classical thermodynamics and computer science. In the next chapter we will lay the groundwork for understanding the concept of quantum erasure as presented in Chapter 9.

Key Points

- The Szilárd single-atom engine uses information, but not a Maxwell demon sorter.
- We use a demon heat sorter and find that waste work is associated with heat rejected to the reservoir in the amount $kT \log 2$.
- The entropy of erasure is simply the ratio of the amount of heat generated (while compressing the atomic packet) divided by the temperature, i.e., an erasure entropy of amount $k \log 2$ is generated on each cycle.

8
Quantum Mechanics II
The wave side of particles and the particle side of waves

Erwin Schrödinger (1887–1961)
Followed the idea of wave-particle duality to discover quantum mechanics.

Werner Heisenberg (1901–1976)
Followed an operational approach to discover quantum mechanics.

"In 1923, de Broglie put forth the following hypothesis: material particles, just like photons, can have a wavelike aspect. Davisson and Germer strikingly confirmed the existence of a wavelike aspect of matter by showing that interference patterns could be obtained with material particles such as electrons or atoms."

Adapted from Cohen-Tannoudji, Diu, and Laloë, *Quantum Mechanics*

The Demon and the Quantum, Second Edition. Robert J. Scully and Marlan O. Scully

ISBN 978-3-527-40983-9

8.1 Erwin Schrödinger, One Father of Wave Mechanics

Quantum theory began when Max Planck tried to understand the properties of light coming from a hot oven. In the last chapter we concluded by noting that the concepts of observation and information erasure are central to modern quantum mechanics. In this and the next chapter we explain these subtle issues in some detail. A key facet of quantum theory is the dual wave–particle nature of matter as conceived by Louis de Broglie and developed into wave or quantum mechanics by Erwin Schrödinger.

As an Austrian, Schrödinger (1887–1961) was born a citizen of the once huge Austro-Hungarian Empire that stretched from modern-day Russia to northern Italy. That was all lost in the "Great War." Erwin Schrödinger came from a well-to-do family that could afford to send their child to a prestigious gymnasium (secondary school). From Einstein and Planck to Heisenberg and Schrödinger, one is hard-pressed to find a modern German scientific giant whose training did not begin early on. Those Europeans who could afford the elite schools were learning Latin, Greek and philosophy. Schrödinger loved philosophy, the never-ending search for truth and insight, and nearly became a philosopher. His work and thinking were often guided by his search for a philosophically consistent picture of reality.

One might wonder why this man, a poet and a romantic, chose the field of science. Perhaps he was directed by his understanding of and appreciation for mathematics. Schrödinger understood that math was the foundation of science and physical reality. As a philosopher, deeply concerned with the nature of reality and consciousness, he saw math as the royal road into the heart of physics. He once told his students,[45] "*first year do nothing but mathematics, second year nothing but mathematics, in the third year you can come and talk with me.*" This dedication to mathematics certainly worked for him. Math, philosophy, an investigative spirit, and Germanic mental discipline were evident in all he did. Schrödinger developed Newton's theory of color into a much more useful scheme. The three primary colors were triangulated and plotted efficiently. By the early 1920s, he was recognized as the world's leading authority on color. World War I brought a break from physics for Schrödinger. As an artillery officer he distinguished himself, demonstrating that he was an excellent leader. He was decorated for a 1915 battle on the Italian front and cited for bravery as follows:

> "In the battle of Oct. 23 to Nov. 13, while acting as a replacement for the battery commander, he commanded the battery with great success ... By his fearlessness and calmness in the face of recurrent, heavy, enemy artillery fire, he gave to the men a shining example of courage and gallantry. It was owing to his personal presence that the gun emplacement always fulfilled its assignment exactly and with success in front of heavy enemy fire ..."

Always the scientist, Schrödinger was investigative even during wartime. He noted that the deafening roar of an artillery blast decreases as one distances oneself from it, but at a distance of 50 to 100 kilometers the noise actually *increases*. Upon his return from the front, Schrödinger wrote a paper inspired by this finding and dealing with the "outer zone of abnormal audibility."

A defeated army seldom receives a welcome home, and while he encountered no hostility, Schrödinger returned to a crushed economy and a bleak future. Wounded and disabled soldiers, who numbered many, had no support, and they often became beggars. Even with careful rationing, starvation was common in Austria. His family's business had been ruined and his father's pension was rendered worthless by the staggering inflation. He did find work as a physicist, but the pay was dismal. Schrödinger would be forever scarred by the condition of his once well-off parents, who were now forced to live at subsistence levels. But conditions eventually brightened, and he developed a reputation in the physics community. Moreover, he found Viennese night life to his liking. And Vienna was a bustling center for the intelligentsia hospitable to philosophers, poets, and scientists. Not surprisingly for a man of his intellectual gifts and adventurous temperament, Schrödinger loved the metropolis.

In several respects, a physicist's career resembles that of an actor's. One needs the first big break to make it in, and that break, a major discovery or a major result, needs to come fairly early in life. Schrödinger was waiting for his major break, but it had yet to come. And the clock was ticking. One reason physicists need their break early is that they seldom make new discoveries when they are older, if they have not begun to do so while young. Furthermore, when they are older they, like everyone else, become set in their beliefs. Revolutionary and exploratory actions are the domain of youth. As such, physicists cut their teeth early in the field if they are to ever really make it. The first decade of Schrödinger's professional life was relatively bland. His early contributions were superior though not ground-breaking. Then in 1925, at the age of 38, Schrödinger gave birth to wave (or quantum) mechanics in a series of brilliant papers.[46] We tell the story of what followed with special emphasis on the experiments that showed the wave–particle duality of matter.

8.2 The Dual Wave–Particle Nature of Light

The debate over the nature of light is centuries old. Newton thought light was a beam of particles. About a hundred years later Thomas Young demonstrated that light consists of waves.

In order to understand the wave properties of light (and matter), let us briefly consider the way water waves behave. Suppose we drop two stones into a quiet pond. Circular waves will propagate out from each point of disturbance. When the waves meet crest to crest the amplitude of the surface disturbance will be doubled. This is called *constructive interference.* When the waves from each stone meet crest to valley, the action on the water will be canceled. This is called *destructive interference.*

The interference between two outgoing waves can also be seen when incoming waves hit two piers or poles in a lake. Interference between waves emanating from two sources is shown in Fig. 8.1.

Another good example of the interference of waves from two sources is provided by the interference between the two tuning forks in Fig. 8.2(a). Here we observe loud (constructive interference) and soft (destructive interference) regions produced by interference from the two tuning forks. Now compare this with the overlapping distribution of particles on a screen from two aerosol spray cans in Fig. 8.2(b). The sound waves show interference behavior in Fig. 8.2(a). The droplets (particles) from two aerosol spray cans show no interference in Fig. 8.2(b).

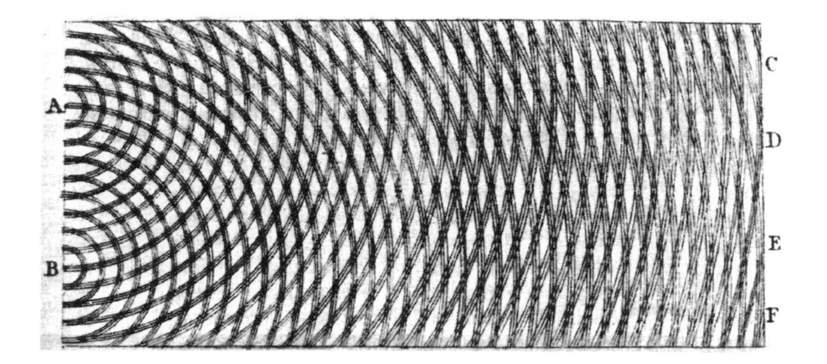

Fig. 8.1 Interference between water waves (be they water, sound or light waves) emanating from points A and B. Constructive and destructive interference is observed on the screen. (Adapted from a figure by Thomas Young.[47])

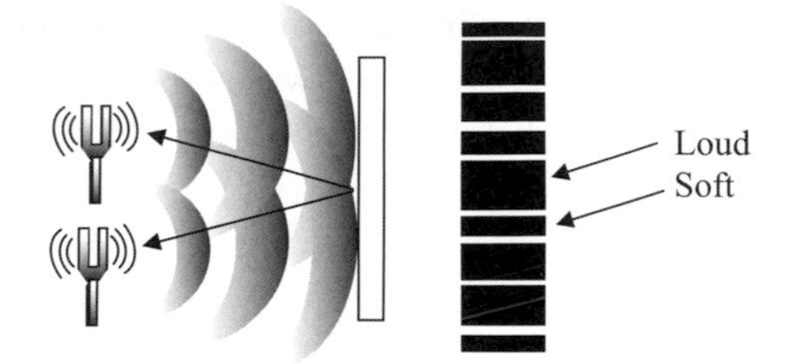

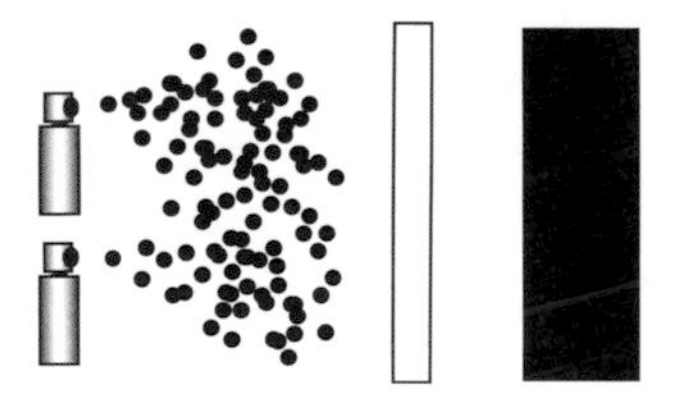

Fig. 8.2 Comparison of the interference pattern produced by two tuning forks (audio wave sources) and the overlapping droplet distribution pattern (no interference) from two aerosol spray cans (particle sources).

The wave nature of light was clearly demonstrated by the two-slit interference experiment of Thomas Young (1773–1829). Young's experiment, a bona fide demonstration of wave optics, was actually quite simple. An experiment using light from a mercury lamp works well. Today's experiments using a laser light source are even easier. The slits have to be positioned very close together. Spacing of the slits at several light wavelengths apart yields a nice light and dark pattern from constructive and destructive interference on the observation screen.

Young got it right. Newton had convinced the world of the particle-like nature of light, so Young's demonstration of light waves was hot stuff. Today, the two-slit experiment is repeated in classrooms around the world.

Great discoverers often rely upon deep intuition. Its role in science can escape attention because the results of the scientist stand on the pillars of sound mathematical logic. Einstein's surprising insights into the workings of reality and existence still astonish us many decades after his death. Young was a master of intuitiveness whose insights into the behavior of light may have been prepared by his precise understanding of the workings of the human eye. Not just a scientist, he was primarily a dedicated physician, known as the founder of physiological optics. He attributed the adjustment in the curvature of the crystalline lens in the eye as being the method for focusing the eye. He not only described astigmatism, but developed a hypothesis that color perception is determined by nerve fibers in the eye which detect red, green, and violet light.

He also employed his sixth sense in a parallel career of Egyptology. Using the Rosetta Stone, with its three different languages detailing the same information, workers had been trying to decipher Egyptian hieroglyphics for years. Guesswork was often the only tool available. Until Young broke the code, his intuition had been at work, backed by tremendous intelligence and knowledge. By the age of fourteen, he was acquainted with 12 languages, among which of course were Greek and Latin. He was a genius and a true renaissance man. He was well versed in every aspect of Western science of the time. When he died in 1829, science was becoming increasingly specialized, yet Young's intuitive nature, perhaps deeper and more subtle than we can understand, was a gift that is most valuable for a scientist.

Young's wave theory of light was universally accepted, with the prediction of Maxwell that light is an electromagnetic wave and its experimental confirmation by Hertz. The wave theory was unchallenged until 1905, when Einstein came up with a particle theory of light, based on his study of the entropy of light, and also explained the photoelectric effect. Scientists were surprised by, and skeptical of, Einstein's arguments for a particle description of light. After all, Newton thought light was a beam of particle-like energy and he was proven wrong by Young and Maxwell. Even Max Planck, the founder of quantum theory and one of Einstein's strongest supporters, thought that he was wrong. After the confirmation of Einstein's theory of the photoelectric effect by an experiment of Robert Millikan in 1915, scientists accepted that light has both a particle *and* wave-like nature. When Einstein received the Nobel Prize in 1922, it was for his particle theory of light and "other contributions to theoretical physics." Einstein's Nobel contribution was the introduction of the concept of wave–particle duality.

8.3 De Broglie, Schrödinger, Davisson, and Germer Demonstrate the Wave Side of Matter

Louis de Broglie made the next key step by theorizing that particles other than those making up light, such as electrons and atoms, might also have a dual wave–particle nature. There is no better way to decide if matter is wave-like than to do an interference experiment for electrons. This is basically what Clinton Davisson and Lester Germer did. The technical details of their experiment are not so important as their conclusion: Electrons show wave-like interference just as light does! No centuries of debate; no big-name nay-sayers slowing down progress, as was the case for Newton's opinion that light was made up of particles. Newton's influence may have slowed the acceptance of the wave aspects of light, but indisputable experiments showed interference behavior. Surely the wave nature of electrons, atoms, and neutrons is even more counter-intuitive than the particle behavior of light. However, the experiment of Davisson and Germer would have convinced everyone.

We say "would have" because Schrödinger had already convinced the world of wave mechanics. For example, he was able to use his wave equation to describe the hydrogen atom in a way that is still taught to every serious student of science. Nevertheless, it is the Davisson–Germer experiment that most clearly demonstrates the wave nature of matter. It will be helpful to pause and explain exactly how such experiments are carried out and are to be understood, as described in Fig. 8.3.

Figure 8.3(a) shows the same kind of experiment that Young used in order to demonstrate wave interference effects in light, but using electrons. Counts, or electron hits, in our detector array

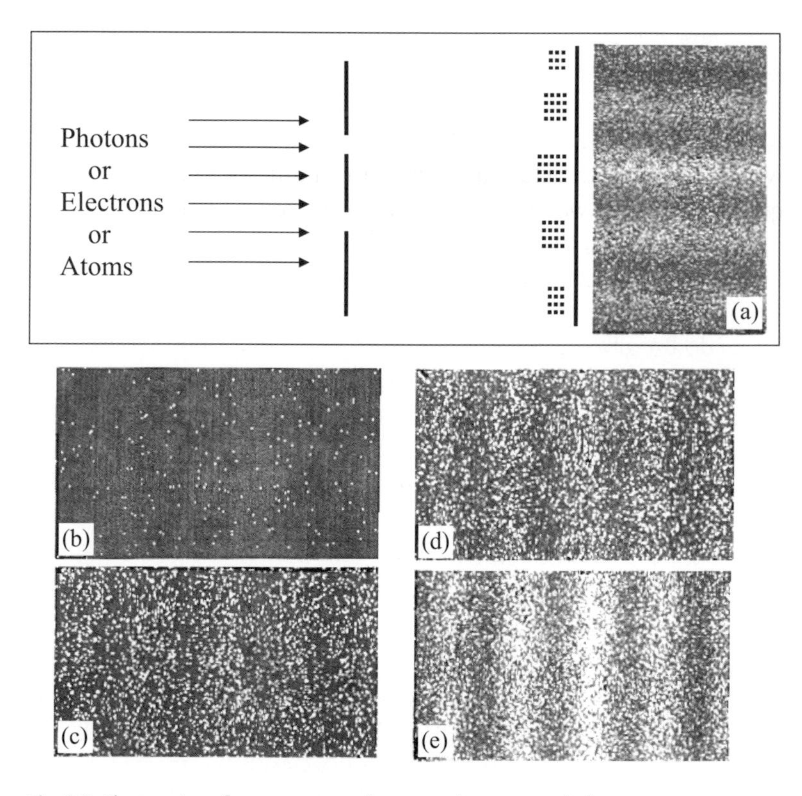

Fig. 8.3 Electron interference pattern from two slits as recorded on film, adapted from material cited in "Further Reading".

(called a film) are counted one hit at a time. That is accomplished by turning the particle flux down so low that there is only one electron or one atom in the air at a time. At first, the counts on the screen appear random (Fig. 8.3b), but as more counts are collected on the screen (Fig. 8.3c–e), an *interference* pattern is seen just like that observed for light!

To summarize, the 1925–26 academic year was to Schrödinger what 1905 was to Einstein: it was a wonderful, massively creative year the likes of which mortals enjoy rarely. It would be interesting to follow all the influences that have led to modern quantum mechanics and the present philosophical interpretations thereof. However, such an account would take us far beyond the scope of this book. Instead, we emphasize that it was Schrödinger who brought the wave mechanics (quantum mechanics) into sharp focus and thereby convinced everyone that it was correct. He did this (before the Davisson–Germer experiments) by developing his famous wave equation. Using the wave equation, he solved the problem of the hydrogen atom in which a single electron is bound to a proton. This yielded a beautiful treatment of the problem, which agreed with actual observation of light spectra, i.e., the frequencies of light waves emitted and absorbed by hydrogen.

Another interesting insight into the development of the Schrödinger equation is provided in an excellent book by Walter Moore. He notes that the last contribution of Schrödinger before his wonderful "explosion of creativity in his discovery of wave mechanics" was an excellent series of lectures and papers on statistical mechanics. Thus it is that Schrödinger brought with him a strong knowledge of statistical physics as he began his odyssey into the quantum wave properties of matter. We discussed this wave characteristic earlier in explaining the interference pattern displayed by electrons passing through two slits.

Schrödinger won the 1933 Nobel Prize for his work in wave mechanics. Finally, he was able to write his own ticket. However, he ran into serious problems with the Nazis, who had taken control of Germany in the same year. Schrödinger loathed the Nazis and everything they stood for. In fact, Schrödinger loathed politics in general. He felt governments of all forms were prone to corruption at all levels and hence were not worthy of man's undivided loyalty or heartfelt trust.

The Nazis were clever and relentless in detecting and pursuing dissent. His true feelings were uncovered in due time. Although Schrödinger foresaw what was coming, he refused to submit to Hitler.[48] Schrödinger fled the country with his family, delivering a public slap in the face to the Nazis.

He found a position in Ireland, where he remained for 17 years. Now the next war was on and rationing began yet again. It was not so bad, at least not until the final days of the war. He had seen much worse. He handled such privations well; to contend with gasoline shortages, he became an avid hiker or he would ride everywhere on a bike. Even when the serious rationing began, it amounted to reasonable, modest cuts, unlike his days in Vienna when the rationing reflected a total collapse of the ability of the system to meet demand. Compared to life on the continent, life in Ireland was peaceful.

A despotism as brutal and hypocritical as the Nazis attempted could not stand for long, and after its fall Schrödinger eventually felt a call to return to his homeland. His Austrian pride had not been diluted by the interlude of fascism. He had left on its arrival and would return on its demise. In 1956, he did repatriate to Vienna and resumed the life he'd always known. Schrödinger died in 1961 and left us with a demonstration that loyalty and greatness come from one's soul. His love of science, philosophy and his fellow man, and his hatred of tyranny mark him as a true Pythagorean. Rather than clinging to the coattails of political fashion, Schrödinger chose to advance humanity by advancing our understanding of reality.

8.4 Werner Heisenberg, the Other Father of Quantum Mechanics

Schrödinger and Heisenberg both received the Nobel Prize for their pioneering work in establishing quantum mechanics. Heisenberg (1901–1976) won it in 1932, a year before Schrödinger. Yet, as described above, they had very different opinions as to the basic foundations of quantum mechanics; not surprisingly, they were strong competitors. Heisenberg focused on the things we measure, like the momentum and position of a particle. Schrödinger concentrated on the state, e.g., wave function, of the particle. Ultimately, their theories were merged to form quantum mechanics. In this section we meet Heisenberg, the man. In the next section we discuss one aspect of his take on quantum mechanics.

As a German citizen living during the war, Heisenberg is also testament to the devastating effects of an intrusive government that allows for no creativity or personal initiative among its populace. Heisenberg's life is also a biography of the dynamics of a people who were exploited to produce one of the most ruthless regimes in history.

Heisenberg grew up as part of a generation of Germans representing the culmination of a century of increasingly resolute nationalism. Their folklore, schooling, philosophy, and literature had grown more and more hostile to foreigners. The Germans had been largely ostracized by the rest of the world community after their defeat in World War I. This, coupled with domestic problems and perceived threats, caused Germany to draw closer together, producing racial "pride."

Heisenberg is a telling illustration of how the young Germans were treated in those times. When only five or six, he was offended by a teacher who abused his trust with an unfair punishment. Even though he was an excellent student, he refused to cooperate for the remainder of the school year.[49] The son of a professor, he excelled in music and saw the keys to nature in its mathematical rhythms. Like many Germans, he loved music, philosophy, life and victory. Toward the end of his schooling he was deliberating between a career in classical music or one in physics. His mother advised him,[50]

> "The future of the world will be decided by you young people. If youth chooses beauty, then there will be more beauty; if it chooses utility, then there will be more useful things."

By that time, Heisenberg was the leader of a troop of Pathfinders. They were Germany's equivalent of Boy Scouts and were a forerunner to the Hitler Youth. If youth are the backbone of their country, these Pathfinder movements were the backbone of the youth. They were all-male groups who bonded, matured, and developed toward their vision of a Germany that would take its rightful place as a superpower.

When Heisenberg's Pathfinders met they hiked and climbed snow-capped mountains, where they dreamt of the Germany they knew in their unique folklore and music. The Germany of that era was a glowing, colorful fantasyland of warriors who would be led from their defeat and troubles by a White Knight. Into glory and immortality they would ride, crushing their enemies and realizing victory while exceeding the limits of normal human capabilities. Heisenberg grew especially attached to his group until they disbanded. But disband they did, as all Pathfinder units came under the direct control of their White Knight shortly after Hitler's rise to power. Soon the White Knight even ordered that all official correspondence be signed with the words "Heil Hitler."[51] Even Heisenberg, Germany's youngest full professor at age twenty six, soon learned that *everyone* had to obey the laws of the new police state.

Heisenberg wanted to continue doing physics and teaching as he had been, but this became difficult under the new regime. One of the Nazis' main goals in education was the control of its teachers and strict regulation of the material they were allowed to disseminate.[52] As professors, Heisenberg and others like him were always under the watchful eye of the Gestapo. Due to several politically incorrect statements he made, Heisenberg was interrogated and spied on at length for some time until fully vindicated by a letter from Heinrich Himmler, coming as a result of a friendship between the mothers of Himmler and Heisenberg.

The letter came just in time. The war office was beginning to clamor for a nuclear weapon. Heisenberg became Hitler's lead man in this Herculean effort. Success should, by all rights, have been theirs. They had the world's largest supply of uranium ore in occupied Czechoslovakia. They had had a head start on the project by several years. Why did they fail to develop the bomb? Actually, they did not just fail; they did not even get close. Not only did they fail to enrich the necessary amount of uranium, but they really didn't even scratch the difficult technological surface of producing the A-bomb-triggering (implosion) mechanism.

One reason they failed affords a good lesson. Countries that harbor a repressive bureaucracy and regulation of education and mistrust of the intelligentsia, seriously hamper their scientific progress. The Nazis persecuted some of their best physicists because they were Jewish. Others, who subscribed to "Jewish physics," such as Einstein's theory of relativity, ran afoul of the Gestapo's good graces. It is no wonder their bomb effort failed![53]

After the war, the allies temporarily interned Heisenberg. His captors hoped for enthusiastic cooperation and afforded him excellent treatment. The interrogations yielded answers similar to what we hear from aspiring nuclear countries today. The Germans claimed they never meant to build a nuclear bomb. They were only trying to develop nuclear power for "peaceful uses."

Naturally there are those who pooh-pooh this notion. However, a fascinating new input on this issue comes from friend Peter Keefe (physicist and patent lawyer). He relates the following "insider story" told to him by his thesis adviser, who was a young colleague of Heisenberg:

> "When I was in graduate school in the physics department of the University of Detroit, my thesis adviser was Gerhard Blass, the department chairman. Gerhard received his Ph.D. from Leipzig University just before the war. On one very special day in 1972 he told me a story about the German bomb project, which he had worked on. Gerhard had the office next to Heisenberg, and he could overhear conversations. One day a physicist was arguing with Heisenberg, 'Why are we not working more quickly on our effort to make a bomb, the Americans are working on it!' the physicist said sternly. Heisenberg replied, 'Look at these calculations; they prove an uncontrolled chain reaction cannot provide energy enough for a bomb. Let the Americans waste their time and money on so futile a project.' Convinced, the physicist (who was an SS implant) reported back to Berlin that the bomb project should be put on a back burner. Blass told me Heisenberg and his colleagues had early in the project a secret meeting and decided Hitler would never get the bomb. So they prepared a set of falsified calculations 'proving' a bomb was impossible, just in case. We can all thank Heisenberg for his good moral sense."
>
> Peter Keefe

8.5 Single-Slit Diffraction and Heisenberg's Uncertainty Relation

We conclude this chapter by using the wave picture of atoms to introduce the very important uncertainty relation of Heisenberg. The Heisenberg uncertainty relation is in many ways the deep basis of quantum mechanics. Its essence is captured in Fig. 8.4. There, we

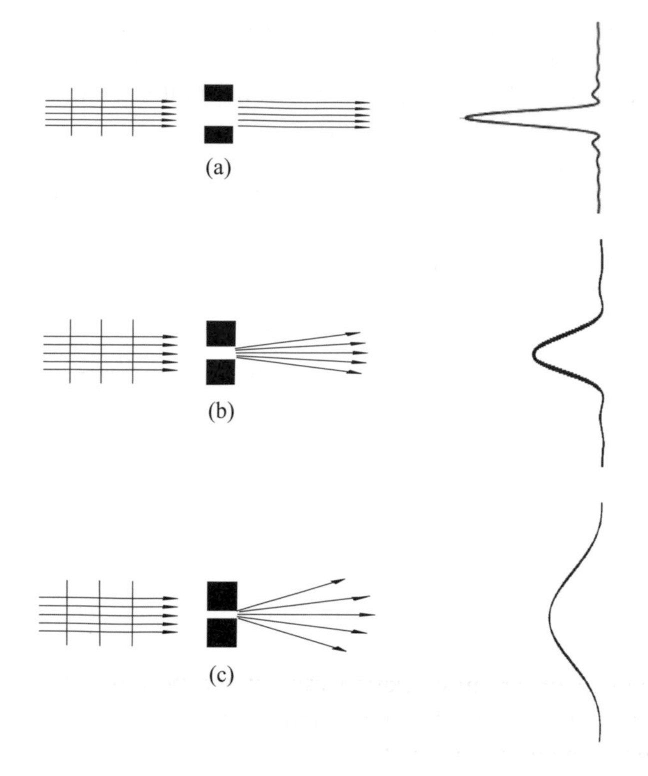

Fig. 8.4 (a) Large slit means little localization (Δx large) but also little diffraction (Δp small). (b) Medium slit gives more localization (Δx small) but also more diffraction (Δp large). (c) Narrow slit width provides more localization (Δx smaller) but now we have much more diffraction (Δp larger).

see that if we are trying to locate a quantum particle, for instance an atom, by accepting only those atoms that pass through a small hole or slit, the more the localization (the smaller the slit) the more the atomic waves are deflected by the slit edges (i.e., diffracted).[54]

The diffraction of particles is essentially a manifestation of the same physics as two-slit interference. This is further discussed in Endnotes 8, and is summarized in Fig. 8.4.

The message of Fig. 8.4 is that the uncertainty in position (Δx) times the uncertainty in momentum (Δp) is approximately a constant, h (Planck's constant). Written symbolically, this reads $\Delta x \Delta p \approx h$. This is the great Heisenberg uncertainty relation. It tells us that the more we try to localize a quantum particle or wave in space, the less we know about its momentum, the product of mass and velocity so the less we know about how fast it is moving through space. The more we know about where we are today (i.e., the narrower the slit), the less we know where we will be tomorrow.[55] We have deflected the particle, because of the narrower slits, and have lost information concerning its velocity in the z direction.

The physics behind the diffraction is essentially the same as the physics of two-slit interference. Think of the single slit as the sum of two slits with zero spacing between them. We should expect to see a (modified) bright–dark pattern on the screen, and this is indeed what we see.

Key Points

- Light and matter both show a wave–particle dual nature.
- The two-slit interference experiment of Young is the archetype demonstration of wave behavior.
- The Heisenberg uncertainty relation says: if the position of a particle is known, its momentum is unknown; or if the momentum is known, the position is not.

In concluding this chapter on quantum mechanics and the men who made it, we include a short vignette on Schrödinger as told to us by Professor Hans Mark whose father was a friend of Schrödinger's.

"Schrödinger and my father, Herman Mark, met for the first time in 1914 when they were in training for the Austrian army. After that, they met occasionally during the war. Schrödinger was a decorated artillery officer. My father won military distinction as a leader of mountain troops. At the end of the war, my father and his men were captured by the Italians, but Schrödinger had left the front earlier for Vienna and so he was not captured.

After the war, Schrödinger was at the university in Vienna and so was my father. They were part of a mountaineering group and would frequently go mountain climbing together. Schrödinger tells the story that they would always argue politics on these climbing expeditions. It was here that Schrödinger would get rather dismissive when he said, "Well, Mark thought I was a socialist." This was "no way true" but this is what Mark thought. My father was kind of right wing.

Another thing that brought them together was the two of them planning to escape after the Nazis' takeover. My father was the one in the most serious danger because his father was Jewish. So they both escaped. Schrödinger to Ireland where he did important work in biology, for example, his little book "What is Life" stimulated James Watson to study biology and win the Nobel Prize with Crick for his famous DNA work. My father took a university position in New York. There he worked as a distinguished polymer chemist on various problems."

The scientific world is a small, but important one. The friendships forged and strengthened by common interests in science are strong and are an important part of the scientific profession. For example, Leo Sizlard, a major hero of our book, took lectures from Professor Mark. Indeed, science is a high calling and the strong bonds between scientific colleagues transcend national boundaries and are an inspiration to us all.

9
From Wigner's Friend to Quantum Erasure

Of Wigner's friends and their amnesia

Eugene Wigner Wigner's friend* John Wheeler

"The quantum eraser ... dramatically underscores the difference between our classical conceptions of time and how quantum processes can unfold in time."

Yakir Aharonov and Suhail Zubairy

Here for the second time in human history subjectivity impacts science. This subjective something (Ψ) is called the *state vector.*

*) This picture was taken at the 1982 Max-Planck Institut für Quantenoptik Workshop organized by Pierre Meystre, Marlan Scully, and Herbert Walther. Wigner's friend is Gershon Kurizki who at that time was a Ph.D. student at the University of New Mexico; and now he is a professor at the Weissman Institute.

The Demon and the Quantum, Second Edition. Robert J. Scully and Marlan O. Scully

ISBN 978-3-527-40983-9

9.1 Introduction

In Chapter 7, we found that the eraser of which-path potential information was an essential feature in resolving the Maxwell demon paradox. In this chapter, we return to the idea of information erasure, this time from the point of view of atomic waves. This will yield fascinating insights into the way nature works; and provide us with a valuable tool to better understand concepts and problems, from Maxwell's demon and Wigner's friend, to wave–particle duality and the famous Einstein–Podolsky–Rosen (EPR) paradox.

In the micro-quantum world the facts of life are altogether weird. They often run counter to what we might call common-sense perceptions about reality and certainly counter to everyday experience. Many of these effects are the result of what Bohr called the principle of complementarity. Wave–particle duality is an example of complementarity. The idea is that once we know the precise value of one of two complementary variables, we can not know anything about the other. For example, if we know which slit an atom goes through in a Young-type experiment, we will not see the expected wave interference fringes. Instead, we will see particle behavior on the impact screen.

In fact, we do not even need to "know" or "read out" the information contained in a measuring apparatus to feel the full force of complementarity. If information on one complementary variable is in principle available, even if we choose not to "know" it (e.g., we choose not to read the meter that contains the information), we still lose the possibility of knowing the precise value of the other complementary variable.

In Chapter 7, we established the fact that Maxwell's intelligent being, or demon, leads to a connection between quantum mechanics, information science and thermal physics. That is, in arriving at a resolution of the demon problem, we found a connection between entropy, informatics, and thermodynamics. A similar connection between complementarity and information in quantum physics will emerge as we proceed.

Eugene Wigner points toward one way of sharpening the information–complementarity interface. He introduces an intelligent "friend" who can make observations without disturbing the object being observed. In this regard, let us return to the wave–particle duality issue. Recall, as we discussed in Chapter 7, that when a sequence of single atoms, or photons, etc., goes through a two-slit apparatus, an interference pattern is observed on the impact screen. Suppose now that Wigner's friend is able to determine which slit the atoms go through. This observation (information) will always rub out the interference fringes. So we ask the question: When is the fringe pattern destroyed? Is it when the friend *makes* the measurement? Or is it only when the information in our friend's mind is *communicated* to us?

> "It is natural to inquire about the situation if one does not make the observation oneself but lets someone else carry it out. What is the wave function if my friend looked ... However, even in the case, in which the observation was carried out by someone else, the typical change in the wave function occurred only when some information (the *yes* or *no* of my friend) entered my consciousness. It follows that the quantum description of objects is influenced by impressions entering my consciousness."
>
> Eugene Wigner[56]

Fig. 9.1 Roy Glauber.

It is important to note that the Wigner friend problem involves essentially two measurements: what do I find, given that my friend found such and such. Thus the outcome of my measurement is said to be "conditioned" upon the results of some other measurement. In fact, the "friend" can be replaced by a simple measurement apparatus, like a photodetector. In modern quantum experiments, scientists often make two measurements involving, for example, two photodetectors or one photodetector and one atom detector. Roy Glauber (Fig. 9.1) received the Nobel Prize for physics in 2005 for developing the theory of such conditional measurements in the context of quantum optics and explaining the statistical properties of light in fully quantum mechanical terms, in addition to many other contributions; for further discussion see Endnotes 9.

9.2 The Quantum Eraser Concept

This observer-participant (Wigner's friend) problem was taken a step further by Marlan Scully and Kai Drühl, who asked: What happens if we erase the information in the mind of the observer? Will the fringes return? The *Newsweek* magazine article reproduced in Fig. 9.2 explains this well.

The *Newsweek* item makes it clear that the complementarity/wave–particle duality aspect of nature is trickier and more subtle than was previously understood. As the article puts it: "No wonder Einstein was confused." When the photons are watched, they behave differently than when not being watched. In order to convey this curious bit of physics, we eavesdrop on a dialog between the interested layman, Man-on-the-Street (MOS), and a learned Professor-of-Physics (POP).

MOS I don't see why there is a big deal about the measurement business in quantum mechanics. We all know that the process of making a measurement or observation is going to disturb the system we look at. For example, when we "look" at an electron by shining light on it, the electron is bumped. There is nothing surprising about that.

A Little Knowledge ...

From 1927 through the 1930s, Albert Einstein fought a friendly battle with Danish physicist Niels Bohr, pointing out the absurdities of quantum mechanics. But Bohr won: every prediction of quantum mechanics is borne out by experiment.

We have long known that ...

PARTICLES OF LIGHT, CALLED PHOTONS, LEAVE A SOURCE ...

AND PASS THROUGH TWO SLITS IN A SCREEN ...

INTERFERE WITH EACH OTHER ...

AND FORM BANDS OF LIGHT AND DARK ON A SCREEN.

Then we found that ...

WHEN PHOTONS ARE EMITTED ONE AT A TIME ...

AND A DETECTOR MONITORS WHICH SLIT EACH PHOTON PASSES THROUGH ...

THE PHOTONS DON'T FORM THE STRIPED PATTERN, INSTEAD MAKING TWO BRIGHT SPOTS.

Somehow knowledge of the paths of the photons through the slits affects their behavior. It's as if they know they're being watched.

Now we discover ...

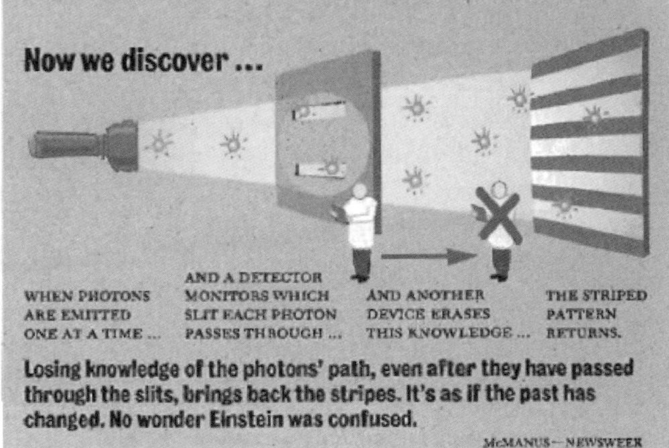

WHEN PHOTONS ARE EMITTED ONE AT A TIME ...

AND A DETECTOR MONITORS WHICH SLIT EACH PHOTON PASSES THROUGH ...

AND ANOTHER DEVICE ERASES THIS KNOWLEDGE ...

THE STRIPED PATTERN RETURNS.

Losing knowledge of the photons' path, even after they have passed through the slits, brings back the stripes. It's as if the past has changed. No wonder Einstein was confused.

McMANUS—NEWSWEEK

Fig. 9.2 *Newsweek* sidebar explaining the quantum eraser concept.

POP I think it is more subtle and much deeper than that. Consider a situation in which we are able to pass particles, say atoms, which have wave-like behavior, through two slits. Then if we do this experiment many, many times, atom by atom, over and over again, we will see on the screen counts in certain positions and no counts at other positions on the screen.

MOS OK, that is the situation with waves that Young explained years ago and you are simply pointing out that the same is true for atoms.

POP Correct, the interference pattern builds up and it is describing a wave-like property of the atoms. But now if I make a measurement to determine which slit the atoms pass through, I lose the interference pattern.

MOS OK, this is just an example of scattering of the atom due to measurement; that, or the uncertainty relation, probably describes all of this quite nicely.

POP Well, you might think so, but we have learned in recent years something new. We are now able to pass the atoms through a particular kind of apparatus that tells us which slit the atoms passed through, but *does not* disturb the atom in any essential way. So we can have "which-path" information and not disturb the atom. Then you would expect that the interference fringes would still be there even though we know we have "which-path" knowledge.

MOS Yes, but there are always these random nudges or shoves applied to the atoms as we "look" at them.

POP Yes, you might think so, but suppose we ignore such pushes and shoves, what do you suppose would happen then?

MOS Well, then I would expect that I would see an interference pattern, but this is very artificial because there will always be such random jiggles.

POP Again, you might think so, but it turns out that your classical intuition has led you astray. We find that the simple *knowledge* of "which-path" information is enough to rub out the interference pattern. Furthermore, we can show that it is possible to make these observations without introducing any random uncertainty of the particle's path.

MOS But isn't that rather artificial? Wouldn't any real detector have to affect the system?

POP Sure, but we can minimize such actions on the system, and we find that we don't even need this back-action to knock out the interference pattern. Knowledge alone rubs out the interference pattern.

MOS Well, this is very strange and it runs against my common-sense.

POP Yes, and equally surprising, if we erase the "which-path" info, the fringes can be regained.

MOS Well I would sure like to understand that.

POP Great! I hope this will leave you feeling happier about these weird, but fascinating, facts of (quantum) life.

9.3 The Quantum Eraser Experiment

To reinforce and simplify the preceding conversation, I will next demonstrate the procedure used in the quantum eraser experiments. I will do this in several steps with diagrams detailing the various stages of the experiment. First, let us look at the ground rules and notation for the quantum eraser experiments as they apply in the following four diagrams (Figs. 9.3–9.6).

1. As in previous chapters, the symbol $\overline{\bullet}$ represents an atom in the ground state, whereas $\underline{\bullet}$ represents an atom in the excited state.
2. When an excited atom passes through a cavity, it emits a photon into the cavity. Thus if we put two cavities in front of the two slits in a Young-type experiment, an excited atom leaves a telltale "which-path" photon in either the upper cavity, associated with passage of the atom through the upper slit, or the lower cavity, associated with passage through the lower slit.
3. The two cavities are separated by shutters and a photon detector. Note that, in the experiment, the excited atom will pass through one of the two cavities (we do not know which), but by opening the shutters we expose the photon to the detector and it can be absorbed (erased) by the detector.

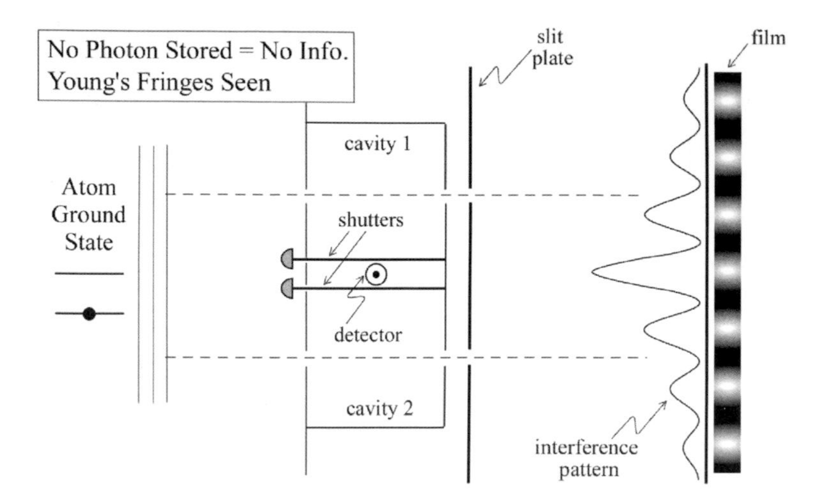

Fig. 9.3 Here we're using a ground-state atom. Thus, it cannot leave a photon in the cavity, so there is no tell-tale photon. Hence, this experiment cannot yield which-path information, and so we have interference fringes on the detection screen. The interference fringes are indicated as a bright–dark pattern (on the film) and the number of counts is sketched as the usual interference pattern.

4. After the atom leaves the cavities and passes through the famous two slits of Thomas Young, we can elect to open one or both shutters. If we open the top shutter and the atom did pass through the top cavity, then we will get a click from the detector when it absorbs the tell-tale photon. The detector is a lot like the film in a camera; it is sensitive to (absorbs) photons. When a photon is absorbed, the detector "clicks." Obviously this which-path click, or event, supplies information (upper path or lower path). Thus the presence of which-path information leads to a rubbing out of the interference fringes. However, if we don't get a click, this means that the atom went through the bottom cavity, which is information all the same. Again, the interference fringes are rubbed out.
5. Contrary to what you might expect, there is no momentum kick or perturbation of the type that one would normally associate with the emission of a photon. That is, there is no perturbation of the motion of the atoms because the kinetic energy of the atoms is very much larger than the photon energy. Nevertheless,

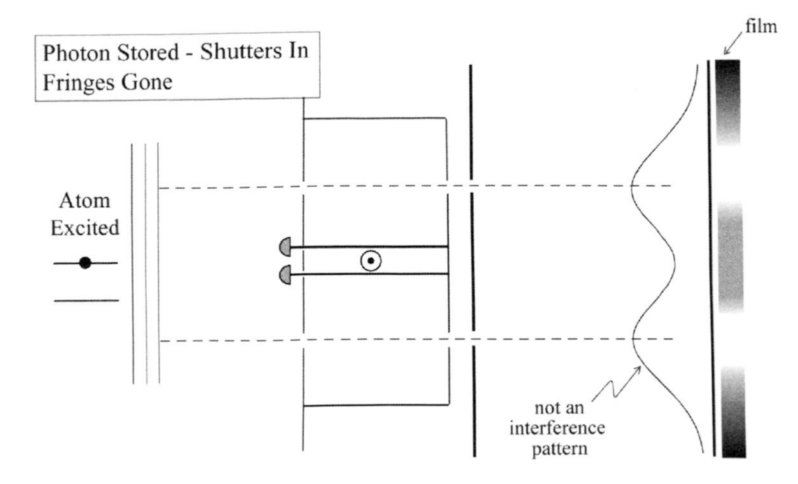

Fig. 9.4 This is the same as Fig. 9.3, except that now we are using an atom in the excited state. Because the atom now leaves a photon in the top or bottom cavity, which-path information is available – if we only open a shutter. But we're not going to in this case. Nevertheless, the *availability* of which-path information rubs out interference fringes. On the detection screen we see two humps – the group of atoms traveling in a straight line from the top cavity gives the top hump, and the group from the bottom cavity gives the bottom hump.

the interference pattern on the screen is indeed rubbed out. The point is that it is information, not perturbation, that rubs out the wave-like interference pattern. One can say that, instead of uncertainty, it is now certainty of which path the atom takes that causes us to lose the interference, wave-like pattern. We could know, if we chose to look, that we went along the top or bottom path according to which cavity has the photon; but we do not even have to look. It is enough just to have the information available.

This loss of interference due to information has been much debated. Much of the resistance to this conclusion arises because it runs contrary to the dogma of conventional quantum mechanics, i.e., it runs counter to the "wisdom" of Bohr, Feynman, and others (as discussed further in Endnotes 9), who frequently argued that it is the uncertainty relation that prevents which-path and wave-like (interference) information at the same time.

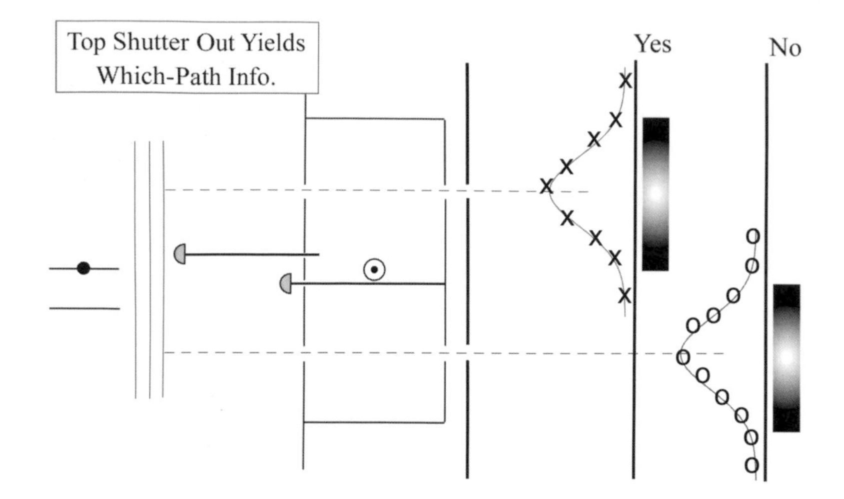

Fig. 9.5 Now we work the shutters. Suppose the top shutter is opened after the atom has passed. Each time we mark the corresponding result on the detection screen. If the photodetector clicked, this means that the atom took the top path and we mark the spot on the screen with a cross. If we get no click, the atom went through the bottom cavity. We mark these spots on the screen with a circle. We have which-path information, and therefore no interference fringes.

6. Now, here is the quantum eraser bonus. Open both shutters, in effect erasing the information we could have had – then what? Will we have fringes again or not? The answer is "yes", as is explained in detail in the caption of Fig. 9.5.

We have explained how it is that erasing "which-path" information can allow us to regain the interference pattern. But, as was emphasized in the *Newsweek* article: "It's as if the past has changed. No wonder Einstein was confused." Some have likewise expressed dismay over this result and have written articles with phrases like "the fallacy of delayed choice and quantum eraser" in their titles. Others, however, have written papers in which the authors maintain that the quantum eraser is: "... one of the most intriguing effects in quantum mechanics." In any case, the situation is summarized in a recent article by Aharonov and Zubairy (see Endnotes 9) which said that the quantum eraser proposed by

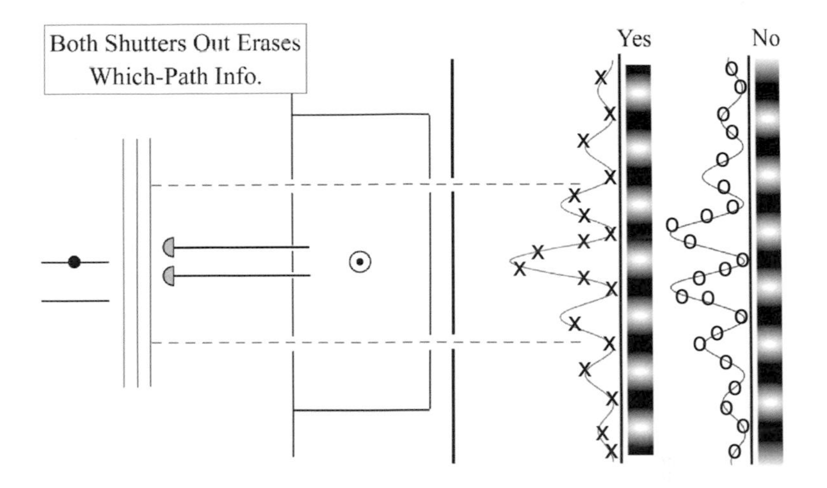

Fig. 9.6 What's different here? We have removed or opened both shutters. In removing or opening both shutters, we have erased potential which-path information (information we could have had). Now consider events on the detection screen or film – the fringes return if we correlate counts on the screen with clicks in the (eraser) detector. If we get a click, the tell-tale photon was erased and we mark the corresponding spot where the atom hits the screen with a cross. If we do not get a click, even though both shutters are out (and this will be the case half of the time), we mark the spot on the screen with a circle. The collection of circles on the screen will be an interference pattern shifted with maxima and minima as indicated.

Scully and Drühl was an idea that "shook the physics community" when it was first published in 1982. So how do we understand this process of something stored in a cavity, a long distance away from the screen, changing the *situation* on the screen?

How to understand this indeed! Well, the quantum eraser effect underscores the statement: information is a physical quantity. That is, information is real and the utilization of information is what the quantum eraser is all about. If we record all of the spots made by the atoms hitting the screen without erasing, then we will not see an interference pattern. However, as discussed in connection with Fig. 9.5, if we mark with a cross the spots on the screen that are associated with a count in the eraser detector, i.e., the "yes" counts, then we will regain the interference pattern (in the crosses). Note that the amplitude of the interference pattern marked with the crosses will never be as large as was observed in the original experiment before we put in "which-path" micromaser detectors because some of the photons in the "which-way" cavities are never absorbed by the eraser detector. Those spots on the screen must be ignored. Only the counts on the screen that are associated with a click in the eraser detector (occurring when we open both shutters) are to be saved. These are the counts that are marked with a cross and which show interference.

9.4 Quantum Eraser: from Maxwell's Demon to Greene's Cosmos

We conclude this chapter by discussing the connections between and the lessons to be learned from the quantum eraser results in the context of the following:

1. The comparison with the Maxwell's demon information eraser of Chapter 7.
2. The comments of James Jeans ascribing more reality to the wave picture than the particle picture of matter, and the opposing view of Fred Wolf in which the particles are real while waves are only a convenient way of thinking.
3. Brian Greene's comments on the quantum eraser.

9.4.1 *Maxwell's Demon and Quantum Eraser*

In Chapter 7 we found that the erasing of which-path potential information was necessary to achieve repetitive operation in closed thermodynamic cycles. Thus the information eraser in that case has somewhat in common with the quantum eraser presented in this chapter. There are essential similarities and differences.

For one thing, the information erased in a Szilárd engine was really known to us. That is, we read the top/bottom indicator of Fig. 5.1 and use that information to extract useful work from a single thermal bath. However, what is erased in our Maxwell demon heat engine is only potential information that we never acquired or used. In our demon-based heat engine we erased the potential for knowing, just as we erased the potential which-path information in order to regain interference fringes in the quantum eraser.

Moreover, the Maxwell demon information eraser scheme of Chapter 7 was irreversible. The isothermal compression necessary to collapse the probability packet requires repeated collision with the hot walls in order to keep the particle's average energy constant, as we carry out isothermal compression. Thus an essential ingredient in that kind of erasure is the irreversible loss of information due to the interaction with the thermal reservoir, i.e., the walls.

The quantum eraser of the present chapter is essentially reversible. That is, if we pass our atom through the two-slit, two-cavity arrangement of Figs. 9.3–9.6 and then reverse the particle before it hits the screen (and before we monkey with the shutters), the atom will go back through the cavities, pick up the deposited photon, and emerge on the left side of the apparatus in the same state it had to begin with.

This is just a long-winded way of saying the present eraser process (before we look, etc.) is reversible. Note that if we tried to run our atom back through the compressor eraser of Chapter 7, we would never expect to regain the initial injected state in which we have two probability packets (top path and bottom path packets).

To sum up, in dealing with Maxwell's demon we stumbled upon the erasure of (potential) which-path information as the way to resolve the paradox, i.e., to preserve the second law. This is just one example of an application of the quantum eraser. There are many others providing new insights into the nature of reality, to wit the next subsection.

9.4.2 Sir James Jeans' Waves, Dr. Quantum's Particles, and Nature's Duality

In his classic book entitled *Physics and Philosophy*, Jeans provides us with many useful lessons. For example, he is clear and convincing in his teaching that the quantum matter waves are "waves of probability" giving a measure of the probability of finding a particle at a given point. However, in his discussion of the relative virtue and corrections of the wave and particle pictures, he says:

> "The two [wave and particle] pictures seem to tell different stories, but we must remember they are not equally trustworthy ... the predictions of the wave picture can not be other than true, whereas those of the particle picture may or may not be true. When there is a conflict, the evidence of the wave picture must be accepted, while we may be sure that the conflict results from some imperfection of the particle picture."

On the other hand the respected author Fred Alan Wolf (a.k.a. Dr. Quantum) in his insightful new book entitled *Dr. Quantum's Little Book of Big Ideas* makes the opposite statement. He says:

> "The 'waves' of quantum physics are ways of thinking. They're not what's going on in the physical world. Particles, particles, particles – that's real in the real world. Waves are a convenience; they're a way of thinking. Waves of possibility. Waves of probability."

So what is a person to think? Here we have two respected authors making diametrically opposite statements. Happily, we can appeal to the quantum eraser experiments and lessons for guidance. Indeed, the lessons we have learned from the quantum eraser teach us that nature is truly dualistic. We get particle or wave behavior equally, depending on what we ask for.

- Do you want to see particle picture (which-path) behavior? Then open the top shutter, as in Fig. 9.5, and correlate the spots on the screen with the yes–no clicks of the erasure photodetector.
- Do you want to see wave picture (which-fringe) behavior? Then open both shutters, as in Fig. 9.6, and correlate spots on the screen with the yes–no clicks of the photodetector.

It's dealer's choice. You pays your money and takes your chances. Surely everything is probabilistic – where the atom lands on the screen is a matter of "chance". However, we interpret the spots as giving wave or particle information depending on what we chose to do with the information deposited in the cavities.

Thus it is the utilization of information, i.e., the correlation of events in our quantum eraser detector and events recorded on the screen, which gives us an interference pattern. In this way, the information that is stored in the cavities (perhaps even long ago) can be viewed as changing the way we interpret the data on the screen today. For this and other reasons, many scientists often regard the quantum eraser effect as puzzling and eyebrow-raising. We could not do better than conclude this chapter with the following quote drawn from Greene's excellent book:[57]

> "These [quantum erasure] experiments are a magnificent affront to our conventional notions of space and time. Something that takes place long after and far away from something else nevertheless is vital to our description of that something else. By any classical common sense reckoning, that's, well, crazy. Of course, that's the point: classical reckoning is the wrong kind of reckoning to use in a quantum universe ... For a few days after I learned of these experiments, I remember feeling elated. I felt I'd been given a glimpse into a veiled side of reality. Common experience – mundane, ordinary, day-to-day activities – suddenly seemed part of a classical charade, hiding the true nature of our quantum world. The world of the everyday suddenly seemed nothing but an inverted magic act, lulling its audience into believing in the usual, familiar conceptions of space and time, while the astonishing truth of quantum reality lay carefully guarded by nature's sleights of hand."
>
> Brian Greene[57]

Key Points

- The dual wave and particle pictures of matter are equally valid and complementary.
- The mere acquisition of which-path information is sufficient to rub out interference fringes. We don't need to "know" which path; we only need to be able to know (if we can choose to ask or look) in order to lose wave-like behavior.
- Quantum erasure allows us to regain wave (or particle!) information long after the atom passes through the which-path cavities and the slits and hits the screen.
- The Maxwell's demon paradox has much in common with the Wigner's friend conundrum.

10

On Quantum Mechanics and the Big Questions

From quantum controversy to Pauli's spiritual complementarity

Albert Einstein (1879–1955)

Niels Bohr (1885–1962)

"God does not play dice with the universe."

Albert Einstein

"Anyone who is not shocked by Quantum Theory has not understood it."

Niels Bohr

"Suppose, for example, that quantum mechanics [leads us to] an unmovable finger, obstinately pointing outside the subject ... to the mind of the observer, to God, or ... would that not be very interesting?"

John Bell[58)]

The Demon and the Quantum, Second Edition. Robert J. Scully and Marlan O. Scully

ISBN 978-3-527-40983-9

"I once asked John Wheeler, one of Einstein's closest colleagues at Princeton, for a fuller explanation of Einstein's hostility to quantum mechanics. Wheeler stated: 'The whole problem was that Einstein did not believe the Genesis story – he never accepted the Big Bang Theory.'"

Henry F. Schaefer, III[59)]

10.1 Introduction: Hindsight–Insight–Foresight

One good way into a difficult problem is to look at it as many ways as possible. For example, when we must make a difficult medical decision, we get multiple opinions. So let us try to clarify these strange ideas of quantum informatics and erasure by listening to the reaction of other scientists. What did/do the experts have to say about all this? Once we better understand what they are trying to tell us, we are in a stronger position to understand the philosophers and theologians. To this end, I will proceed as follows:

1. First, let us go back to the notion of rubbing out the interference pattern by *disturbing* the particles (atoms, photons, or whatever). How do people (e.g., Bohr and Feynman) usually explain the loss of wave-like interference this way? We will give special attention to the application of Heisenberg's uncertainty relation to this problem. We will be especially interested in the way that Bohr used the uncertainty in position and momentum to advantage in his famous arguments with Einstein. The reprint by Mark Buchanan in Endnotes 10 develops these ideas further following Heisenberg and his discussion of how looking at an electron (by scattering light off it) disturbs it in accord with his uncertainty relation. The contrast with the present (quantum eraser) "it's all information" perspective is thus made clearer.
2. Then we will proceed to review the reaction of various scientists and thinkers to this new "information replaces ordinary pushes and shoves" point of view. This is, after all, how science progresses. Someone puts up a new theory or idea and others agree or disagree. Those who disagree try to shoot the original ideas down. Then, when the various camps have made their positions clear, the experimentalists take over and decide the matter.
3. Having strengthened our understanding of the uncertainty relation and quantum information perspectives via the preceding,

we will consider various theosophical issues. For example, it is sometimes argued that the uncertainty relation allows for free will as contrasted to the classical mechanical predeterminism of Laplace. After reviewing this notion, we ask: How does the present information perspective handle the free will problem?

4. Next we return to the quantum eraser problem and present some interesting thoughts of various scholars about, and arguments against, quantum eraser. Experimental adjudication will then be sketched, with more details given in Endnotes 10.
5. One useful result of the quantum eraser is the way it sheds light on older problems like the Einstein–Podolsky–Rosen (EPR) paradox. The wonderful EPR problem or paradox was really the birthplace of quantum informatics. It is thus interesting to see how the present quantum eraser effect can help us to better understand the old EPR chestnut.
6. With the quantum eraser and EPR problems under control, we make bold to ask: What do people mean when they talk of proceeding from complementarity to consciousness? More to the point: What do the philosophers and quantum physicists who have thought hard about these problems want to teach us? How do we understand their ideas in view of what we have learned about entropy and information?
7. The title of a recent talk entitled "The convergence of science and religion" by superhero Charles Townes is an apt summary of what many quantum mechanics seem to be saying these days. In this, the last section, we will follow the metaphysical and theosophical thoughts of scholars from Plato, Pauli and Schrödinger to Polkinghorne, who teaches that the soul is the sum total of our thoughts stored in the mind of God.

10.2 Quantum Mechanics Has Deflected Many Attacks by Using the Heisenberg Uncertainty Relation

"If an apparatus is capable of determining which hole the electron goes through, it cannot be so delicate that it does not disturb the pattern in an essential way. No one has ever found (or even thought of) a way around the uncertainty principle. So we must assume that it describes a basic characteristic of Nature."

Richard Feynman[60)]

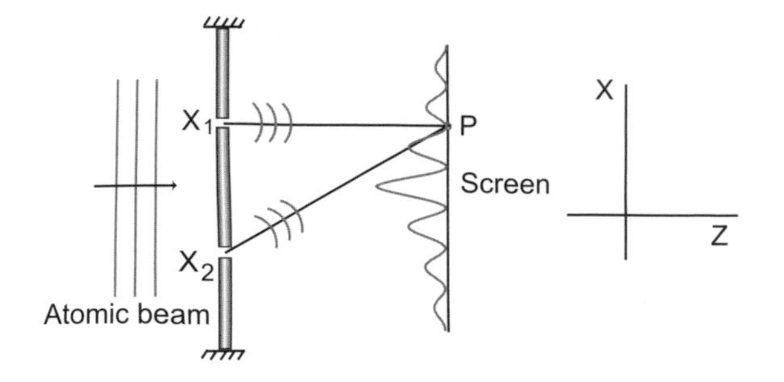

Fig. 10.1 Depiction of atoms passing through double slits X_1 and X_2, which are rigidly fixed. When we collect counts on the screen, one by one, the pattern will show wave-like behavior as in Fig. 8.3 of Chapter 8.

Traditionally, it has been the Heisenberg relation between the uncertainty in particle position and momentum, $\Delta x \Delta p \approx h$ (see Section 8.4), that has been used to protect quantum mechanics against schemes that would seem to violate complementarity and/or wave–particle duality. In this sense, Feynman was correct in saying that (at the time of his writing) no one had ever thought of a way around the uncertainty relation in this regard. However, Scully and Drühl and the team of Englert, Schwinger, Scully, and Walther (ESSW) did "think of" a way around it, namely quantum erasure.

Let us recall, for example, the famous Einstein–Bohr debate involving a recoiling slit arrangement, as per Fig. 10.1. Einstein wanted to show how to violate the quantum wave–particle complementarity by finding out which slit an atom went through and then, one by one, collecting counts and providing a wave interference pattern. He proposed to do this by freeing the slit plate of Fig. 10.1 so that it can recoil.

As depicted in Fig. 10.2, and explained in the caption, the direction of recoil of the slits tells us which hole the atom passed through. For the example of Fig. 10.2(a), the atom passed through the upper hole and travels straight ahead to point P. It therefore experiences no deflecting kick and the slits do not recoil. But if it passed through the bottom hole (Fig. 10.2b), the atom acquires a momentum kick in the positive $\hat{x}$ direction and the slits recoil in the minus $\hat{x}$ direction. Hence, if we record the position of atom

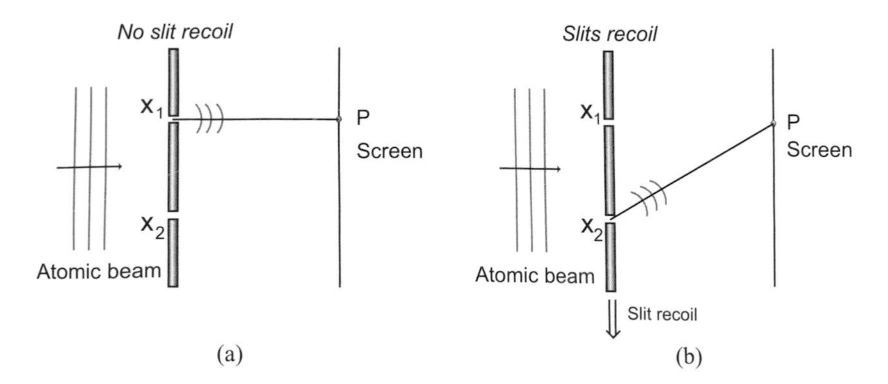

Fig. 10.2 A beam of atoms propagating in the *x* direction (an incident plane wave) is deflected by passing through slits X_1 and/or X_2. Conservation of momentum requires the slit plate to recoil by an amount that depends on where the atom hits the screen and which slit the atom passed through. (a) If the atom is (not) scattered by slit X_1 and hits the screen straight ahead, there will be no recoil of the slit plate. (b) But if it is scattered from slit X_2 and hits the same position on the screen, the slits will recoil.

impact on the screen and the magnitude of the recoil of the slits, we would seem to have both which-path information (by recording the recoil of the slit plate) and observe wave behavior (by collecting many events), thus contradicting complementarity, as explained in Chapter 8.

Einstein presented this clever attempt to confound quantum mechanics at the famous 1927 Solvay conference. Bohr worried all night about Einstein's scheme and came back the next day with a resolution. Bohr explained that the interference pattern obtained by collecting many counts is wiped out since the exact positions of the slits are unknown due to the uncertainty relation (see Fig. 10.3).

In view of the preceding, and other similar examples, it has been generally held that the uncertainty relation is the bodyguard of quantum mechanics. The statement of Feynman, given at the beginning of this section, is a case in point. So the paradox of complementarity still holds true; if we have knowledge of which path, we lose interference fringes. The preceding diagrams and arguments have demonstrated how Bohr and Feynman used the uncer-

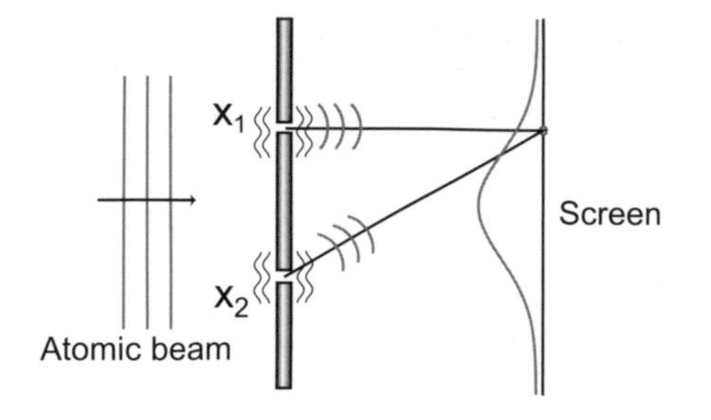

$\Delta x = \Delta p/\hbar$ *uncertainty*

Fig. 10.3 Bohr argues that, if the slit plate is so light as to be deflected by a single atom, then it should be subject to the rules of quantum mechanics. In particular, the slit position is uncertain due to the Heisenberg uncertainty principle. Then if the quantum mechanical uncertainty of the slits is taken into account, the interference fringes will wash out due to various contributions to the interference pattern coming from uncertain points of origin.

tainty relation to protect quantum mechanics. Although they did not fail to enlighten us, their arguments are not complete and can lead to wrong conclusions. For, as we'll continue to see, whether in wave–particle duality, computers, engine design, philosophy, etc., *information* is the lifeblood of science and nature.

Note, as discussed at the beginning of this section, that the quantum eraser scheme of Chapter 9 does show "a way around the uncertainty relation." We expand on this strong statement in the next section by outlining some of the controversy it provoked; see also the reprint by Buchanan in Endnotes 10.

10.3 Uncertainty Versus Certainty: or Heisenberg Meets Entanglement

One interesting argument concerning the quantum eraser scheme of Chapter 9 was given by the excellent team of New Zealand scientists Storey, Tan, Collett, and Walls. They purport to show that it is the uncertainty relation that protects quantum mechanics and preserves wave–particle duality, i.e., protects against simultaneous knowledge of wave and particle information:

"The enforcement of the principle of complementarity in a two-slit interferometer is thus a direct consequence of the exchange of momentum between particle and detector, in an amount satisfying the requirements of the uncertainty principle."

Storey, Tan, Collett, and Walls

In our quantum eraser scheme of Chapter 9, it is essential that the which-way knowledge is acquired without changing the atomic center-of-mass wave function of the atom noticeably – in marked contrast to Einstein's recoiling-slit experiment. This property of our quantum optical which-way detectors has been questioned by Storey and coworkers:

"Taking the proposal of Scully, Englert and Walther [SEW] as an example, we show that in any path detection scheme involving a fixed double-slit, complementarity is enforced by means of momentum transfer in accordance with Heisenberg's uncertainty principle."

In other words, Storey *et al.* contend that it is the Heisenberg uncertainty relation that rubs out the which-way (which-slit) information after all.

The SEW group fight back. They purport to

"... show that the transverse-motion objections raised by Storey et al. are invalid for several reasons: they use an inappropriate set-up of the fields and cavities; they do not finish the calculation. The complete calculation confirms our original statement that it is the presence of which-way detectors, not Heisenberg's uncertainty relation, that enforces complementarity."

To settle the matter, a beautiful experiment was carried out by the German group under Gerhard Rempe.[61] They did indeed find that, when the "which-path" information was available, the interference pattern was lost. As Rempe puts it:

"Everyone believes that when an interference pattern is lost [due to which-path measurements], it happens because a measuring device delivers random kicks to the particles. But there are no random kicks in our experiment."

The bottom line is that Rempe concludes that it is observation, or acquisition of information, that rubs out interference. Thus, as we said earlier, in the quantum eraser experiment, it is certainty, not uncertainty, that rubs out the interference pattern. To put it another way, the process of measurement involves correlating or entangling the particles we are observing with the measuring apparatus. In the words of Rempe:[61]

> "To explain the wave–particle duality, we need entanglement and correlation. The Heisenberg uncertainty relation has nothing to do with wave–particle duality."

Two interesting popular press discussions of the Rempe experiment serve to further clarify matters. P. Weiss in his *Science News* article says:[62]

> "German experimenters find that the mere existence of information about an entity's path causes its wave nature to disappear. They offer experimental evidence that something deeper than uncertainty yanks off stage the wave – or particle – half of the duet. That unseen hand is known as entanglement, or correlations."

The other article, entitled "An end to uncertainty," by M. Buchanan, is included in Endnotes 10.

10.4 Quantum Mechanics, Free Will, and Conscious Choice

10.4.1 *Laplace and Predestination*

The following is an imagined dialog, but basically true to the mindsets of the participants. It sets the stage for this section.

Laplace	Once the position and momentum of every piece of the universe is known, the laws of physics take over and the future is determined.
Napoleon	Where then, in your philosophy, is there room for God?
Laplace	Sir, I have no need for that hypothesis.
Lagrange	Ah, but it is such a good hypothesis, it explains so many things!"
Margenau	[who lived a couple of centuries later] But you see, with the discovery of quantum mechanics; room has been made for decision and choice, which had no place in the older scheme of things."

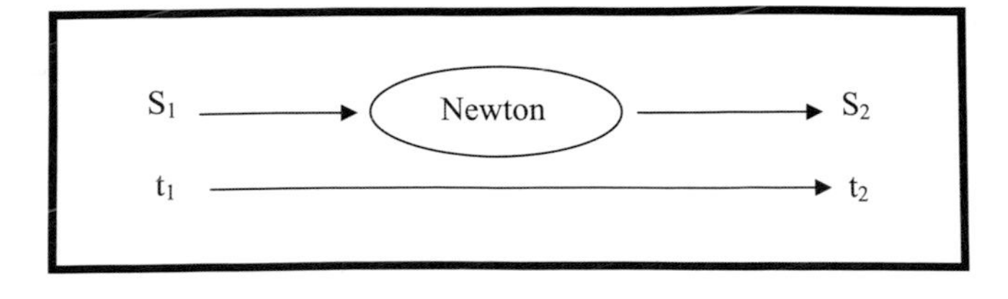

Fig. 10.4 The time evolution of a classical system.

Let us follow Professor Margenau further.[63] In his book *The Miracle of Existence* he teaches us that in classical mechanics causality was a relatively simple notion. This is well illustrated by Newton's laws of motion. Consider the (continuous) motion of any object. At a given time (t_1) it has a position (x_1) and a velocity (v_1); at a later time (t_2) it is at x_2 and has velocity v_2. The two observables, x and v, define the state of the moving object. Given the initial state $S_1(x_1, v_1)$, the later state $S_2(x_2, v_2)$ at the time t_2 can be calculated from Newton's laws of motion. Margenau presents the situation schematically as in Fig. 10.4.

The initial state of the system, S_1, is "fed" into a mathematical formalism called a *law*. Then, by formal math, we can predict the state S_2. That is, if the state of a physical system is known at a given time, then the state at a later time can be computed and thus predicted. We call the earlier state the *cause* and the later state the *effect*. Given the initial condition, everything is determined forever!

Evan Walker, in his thought-provoking book *The Physics of Consciousness*, puts it this way:[64]

> "Classical physics would demand that nature grind out blindly and automatically the consequences of any initial action. Any mind attached to such an automation would be only a passive observer. Such a mind would not be able to control any aspect of its body's behavior. It would be a captive bird in the brain cage, and there would be nothing to call 'will'."

How sad, how very, very sad.

Finally, let us ask Jim Baggott for a one-sentence statement of the "free will" problem. He gives us this in his excellent book *Beyond Measure:*[65]

"Classical physics paints a picture of the universe in which we are nothing but fairly irrelevant cogs in the grand machinery of the cosmos."

"Okay, okay!" we shout. "Classical physics just makes us robots, but how does quantum physics get us out of this bind? We believe in free will and conscious choice; please show us how quantum thinking solves the problem."

To that end, we let our quantum teacher, Margenau, explain his views:

"[Quantum mechanics] makes possible the freedom of our will, which would contradict science if strict determinism regarding the future prevailed. One conclusion we draw from this is that quantum mechanics, or any basic formulation of physical principles, must operate with probabilities, and that efforts to remove them are philosophically misguided."

The validity of his remarks concerning probability is made clear by recalling our treatment of the Young double-slit experiment given in Chapter 8. In particular, Fig. 8.3 shows how any given atom or electron passes through the two slits (in some sense) and then, with a likelihood or probability given by the laws of quantum mechanics, hits the screen at a given point. Only after many particles have passed through the holes and are collected on the screen does an interference pattern emerge.

10.4.2 The Heisenberg Uncertainty Relation and the Free Will Problem

As discussed in the preceding section, classical Newtonian physics would seem to leave little room for free will. There we reported the arguments of various scholars who thought that the quantum description of nature was enough to restore free will. In his book on free will, Harkavy supports this point of view; there we find the statement:[66]

"But if we accept freedom as a fact, we are bound to consider whether at least a certain measure of physical indeterminism may not also be a fact."

However, not everyone agrees. As recorded by Harkavy, whose presentation we follow, the physicist Charles G. Darwin (grandson of the biologist Charles Darwin) disagreed. Nobel Prize physicist A. H. Compton summarizes Darwin:[67]

> "Darwin's argument is that 'physical theory confidently predicts that the millions of millions of electrons concerned in matter-in-bulk will behave ... regularly, and that to find a case of noticeable departure from the average we should have to wait ... longer than the estimated age of the universe.'"

That is, according to Darwin, quantum mechanics applies (in practice) only to the quantum world. Compton disagrees:

> "He [Darwin] apparently overlooks the fact that there is a type of large-scale event which is erratic because of the very irregularities with which the uncertainty principle is concerned. I refer to those events which depend at some stage upon the outcome of a small-scale event."

It does indeed seem strange for the physicist Darwin to take such a hard-and-fast position. It is not difficult to point to current experiments or situations which would require the application of quantum mechanics (e.g., the Heisenberg uncertainty relation), to macroscopic systems such as superconductors, superfluid helium, and laser-cooled Bose–Einstein condensates. Clearly, Darwin is too restrictive in claiming that the uncertainty relation applies only to the micro-realm.

However, we don't need Heisenberg's input to see that quantum mechanics breaks the classical mechanical embargo against free will. Indeed, the quantum eraser lesson shows us that certainty (knowledge) can do the trick, i.e., break the classical death lock on free will. As we will discuss further in Section 10.4.4, the mere presence of an external observer (God) who knows (but never tells) the complete state of our existence (self, soul, and surroundings,...) is enough to ensure an indeterministic time evolution. That is, complete certainty of our total (local) universe projected onto our more limited (less than total) subspace allows for free will. Again: it is certainty, in the mind of God, that injects uncertainty and free will into the soul of man.

10.4.3 *Free Will and Classical Chaos*

But do we really need quantum mechanics to ensure free will? From the perspective of Darwin, the physicist, as presented in Section 10.4.2, it would seem that quantum mechanics might be sufficient, but is it necessary?

Laplace would say that all is (pre)ordained when the initial conditions were set up. We must act our way through life, but we really have no choice in the decisions we make.

"Not so," proclaims the modern classical mechanic! We know that mathematical chaos (i.e., extreme sensitivity to boundary conditions) breaks Laplacian predestination. The new paradigm is the "a hurricane in Florida this fall could be caused by a butterfly flapping its wings in South America last spring" story we have all heard many times in recent years.

But is classical chaos enough to provide freedom of choice, i.e., free will? Surely present-day mathematics and digital computers are limited by this extreme sensitivity to boundary conditions. But perhaps we will learn how to make far better computers (quantum computers?) which are not confounded in the way today's computers are. Perhaps we will come up with new mathematics that will not be so sensitive to initial conditions.

Who knows? Nobody. In an excellent recent *Physics Today* article, George Ellis, famous quantum physicist and philosopher, addressed this question. He is too smart to put forth a dogmatic answer. However, at a conference in honor of Charles Townes' 90th birthday, he was asked point blank: "Do we need quantum mechanics to ensure free will?" His answer:

> "On Monday, Wednesday, and Friday, I think not.
> On Tuesday, Thursday, and Saturday, I think so."

In either case, Laplace would surely be interested to see his predestination doctrine overruled. However, if he would not, surely Napoleon and Lagrange would!

10.4.4 Theosophical Questions Suggested by Uncertainty Versus Entanglement and Which-Way Detectors

In Section 10.4.2, we discussed the way entanglement and informatics replaces uncertainty and noise as the protector of quantum mechanics. What then, can we say of quantum entanglement and the free will question?

Not much has been written about this interesting question; the fascinating paper by Zoeller-Greer, quoted later in this chapter, is a step in the right direction. He makes a clear, strong statement concerning free will and quantum mechanics:

> "Now, with quantum mechanics, God can ... enable human beings to make 'true' decisions. These decisions are not determined in advance by the current state of the universe. So the old dilemma of living in a calculable universe and having a free will is also solved (although God is, of course, omniscient concerning all events that occur in the universe)."

As is further discussed in the following, the presence of an external observer breaks the casual Schrödinger dynamical evolution. That is, if we start our system off in state Ψ_1 at position x_1 and time t_1, now we replace the classical time evolution of Fig. 10.4 by the quantum time evolution depicted in Fig. 10.5. So every Ψ_1 goes over to a specific Ψ_2 in a unique fashion.

Now we may ask what happens when we include an external observer in the problem. What happens then to the time evolution of our state Ψ_1 (which does not include the observer)? The answer is that Ψ_1 no longer goes over to a specific pure state Ψ_2. Instead, we find that additional quantum probabilistic features come into play.

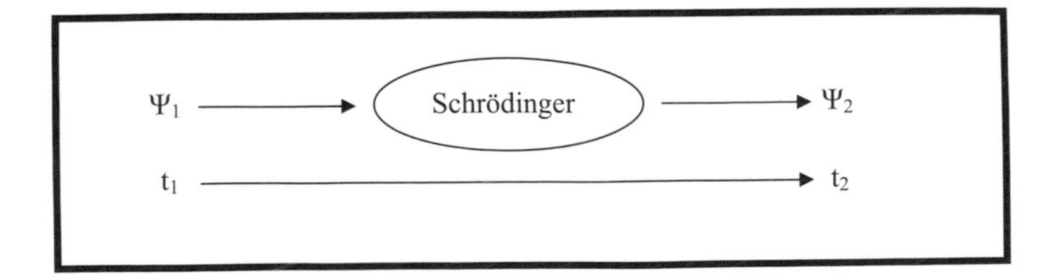

Fig. 10.5 The time evolution of a quantum system.

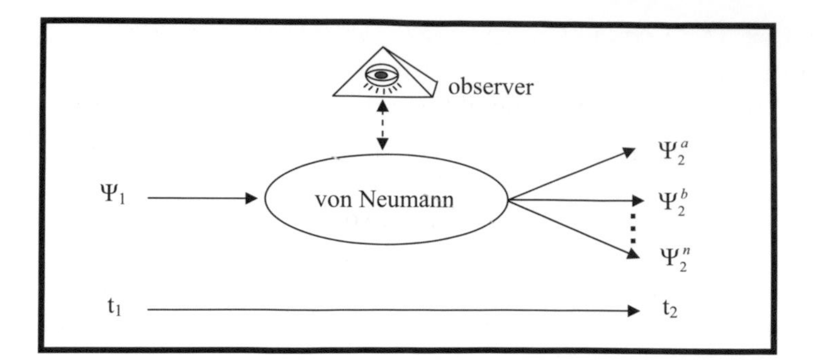

Fig. 10.6 The time evolution of the system is influenced by an external observer so that the states $\Psi_2{}^n$ are connected to Ψ_1 in a *probabilistic* fashion. John von Neumann wrote helpful papers on this problem.

Specifically, the final state is now describable only in terms of probabilities of ending in some state Ψ_2, as per Fig. 10.6.

Clearly, the presence of an external "All-seeing Eye" changes the way we view the object of interest. The dynamical evolution of Fig. 10.5 is replaced by the less well determined probabilistic one in Fig. 10.6. This is the quantum eraser message of Chapter 9; the mere presence of an observer is sufficient to wash out the fringes.

10.5 Quantum Erasure Revisited

10.5.1 Szilárd, Wigner, and Friends: the Hungarian "Kugel Köpfe"

There is a clear connection between the present notion of quantum erasure as inspired by Wigner's friend, the idea introduced in Chapter 5, and that associated with Szilárd's version of Maxwell's demon problem. In Chapter 9, we were erasing "which-path" (top hole–bottom hole) information in order to regain wave-like interference behavior. In the case of the Stern–Gerlach demon engine in Chapter 7, we had to erase which-path (upper path–lower path) information in order to complete the thermodynamic cycle. There is no easy street leading to nature's secrets, but here we have an interesting connection between thermodynamic and quantum erasure. It is also interesting to note that three people whose work provides profound insights into such problems were from the same excellent Budapest High School system.

But let us pause to tell a bit of the story of four Hungarian scientists, called "Kugel Köpfe" (ball-bearing heads) because they were so smart they seemed to think without friction: Leo Szilárd, Eugene Wigner, Edward Teller, and John von Neumann. They easily mastered the trickier concepts in quantum and classical physics, possibly as a result of the outstanding training they received when they were young. The four physicists, in their maturity, developed many new scientific ideas in fields ranging from thermodynamics to quantum mechanics and information (computer) science. They also provided beacons that still light the path of scientific development. This is noteworthy, since, as we have seen, like history, science sometimes repeats itself. In this case, that repetition is to be seen in a comparison between these four men and the Pythagoreans.

Brilliant and industrious, the four "modern-day Pythagoreans" shared another common bond: they were all Jews. They were persecuted during the scourge of anti-Semitism that gripped Eastern Europe after World War I. Like the Pythagoreans, the Jews who fell under Hitler's heavy hand were forced from their homes, stripped of all titles, posts and property. Our "Kugel Köpfe" were scattered, but, unlike the Pythagoreans, in their dispersion they made their biggest contributions to humanity. The four modern-day Pythagoreans, von Neumann, Szilárd, Teller, and Wigner, were the fathers of the atomic bomb, the electronic computer, much of modern quantum mechanics, and more.

Building the bomb required the ultimate intellectual achievement in engineering science that the world had known. It was the biggest scientific project ever, to be accomplished in absolute secrecy in a highly compressed interval of time. The top leaders of that endeavor were some of the smartest men in the world. As a child, John von Neumann would entertain guests at his house by memorizing an entire column in a phone book that had been selected at random. He'd look over the column a few times and then hand the book back to the guest, who would then grill the young von Neumann as to who had which address or what was so-and-so's phone number. As a top mathematician, von Neumann applied his genius to quantum mechanics by writing one of the first books on the subject, and to the development of the atomic bomb by utilizing the implosion physics that he learned when he was a consultant at the Aberdeen Army Proving Ground. After the war, he helped in making the hydrogen bomb. He was one of the chief designers of the first electronic computer. All of this he did by employing his understanding of pure and applied mathematics.

Leo Szilárd, our thermodynamics hero, was another genius. He had an early patent on nuclear chain reaction devices, e.g., the nuclear reactor. He even took out a patent with Albert Einstein, in which they designed a refrigerator with no moving parts. The refrigerator was the perfect example of applying the mathematical side of thermodynamics developed by Carnot and Maxwell. Now those applications were finding their way into the market. In these studies, Szilárd branched from math into engineering and physics.[68]

Edward Teller (1908–2003) was another physicist who contributed to the development of the atomic fission bomb. He also played a key role in the hydrogen fusion bomb project and was a highly sought after defense scientist. As with his three cohorts, Teller became a recipient of many honors.[69] And just like his colleagues, he became an active voice in nuclear deterrence during the Cold War. All four physicists favored a strong deterrent in the form of advanced nuclear weaponry; but they also advocated international, open communication between nations, and UN-mandated control over nuclear weapons proliferation.

Eugene Wigner was a leading mathematical and theoretical physicist[70] who became one of the top dogs in the field of quantum mechanics and won the Nobel Prize. In 1939, Szilárd, Wigner, and Einstein advised President Roosevelt that a nuclear bomb was a real possibility. They told him that scientists in Germany were working on one and that it was imperative that the Americans build one before the Germans did. Thus began the Manhattan Project.

The strong connection between math and science is the hallmark of our "fabulous four." The intellectual background of "Pythagoreans" ancient and modern has always been, and always will be, mathematics. Math is the curiously perfect form of information and communication governing the laws of the universe. As such, it is the key to scientific learning. A young scientist wishing to advance does well to remember this. Technology depends on science and mathematics. Technological strength roughly equals economic strength, a formula to be noted by nations attempting to strengthen their position in the world. As stated by Napoleon:[71] "The advancement and perfection of mathematics are ultimately connected with the prosperity of the state."

The history of the Pythagoreans also sheds light on how necessary it is to have free-thinking teachers. Highly controlled and regulated education is *not* the trademark of a society that hopes to foster and nurture gifted minds. One way to encourage and foster creativity is to provide teachers like the ones in Budapest or ancient Greece. A single, motivated educator can be tremendously effective.

One of the most important lessons Pythagoreans, old or new, have to teach us is the dramatic loss suffered by the society that persists in devaluating scientific research. The original Pythagoreans could likely have altered the fate of humanity if left alone. And these later ones of whom we now speak? If Hitler had left *them* alone, our own fate could certainly have been altered. Obviously, as an open democratic society, we valued and encouraged research at one time. However, when we deviate too far from that prudence, we will repeat history in a manner we should desire to avoid. There have always been Pythagoreans out there – we need only to nurture and listen to them.

10.5.2 *Trying to Erase the Quantum Eraser: Quantum Controversy*

"Englert, Scully, and Walther (ESW) have proposed a two-slit experiment with atoms in which (so they appear to claim) the experimenter can decide, long after an atom has made its mark on the screen, whether that atom has passed through a particular slit or has, in a sense, passed simultaneously through both of them ... The purpose of this paper is to show that, in fact, the experimenter does not have the above-mentioned choice."

Ulrich Mohrhoff

As in the case of the Storey *et al.* uncertainty–certainty debate of Section 10.3, insight into the quantum eraser concept can be gleaned from its critics. One scholar in particular presented a useful criticism. In an article entitled "Restoration of interference and the fallacy of delayed choice: concerning an experiment proposed by Englert, Scully, and Walther" Ulrich Mohrhoff argued that the Young's two slits plus two which-way detector experiment:

> "... proposed by Englert, Scully, and Walther, appears to permit experimenters to choose, after each atom has made its mark on the screen, whether the atom has passed through a particular slit or has, in some sense, passed through both of them ... In actual fact, this choice exists only until the atom hits the screen."

Mohrhoff then goes on to present his reasons for coming to this contra-quantum eraser conclusion. However, he changes his mind (in a later paper) and retracts his criticism, as follows:

> "In a recent article I analyzed the thought experiment of Englert, Scully, and Walther (ESW) from two 'metaphysical' perspectives, the reality-of-states view and the reality-of-phenomena view. In that article I arrived at a wrong conclusion, for which I wish to express my sincere apologies to the readers of this journal. I compounded my mistake by attributing my views to Englert, Scully, and Walther. My apologies also to the authors!"

In fact, Mohrhoff has done us a valuable service by provoking ESW to clarify their views and, in so doing, making contact with the famous EPR paradox. In particular, in their response to Mohrhoff, ESW say (and see following discussion and Endnotes 10):

> "To put the problem in a different light – all of the above is at the root of the EPR problem (not paradox!). Once we have understood one (the preceding, for example) we understand the other (that is, EPR). Some years ago one of us (MOS) analyzed the EPR problem in exactly this spirit, and we present a brief review of that analysis."

They then go on to present their analysis of the EPR "problem" as they call it. Plenty of other people call it a "paradox," but never mind. The details of the ESW analysis is given in Endnotes 10. The main point is their conclusion as contained in the following quote from their rebuttal paper (which is translated to the present quantum eraser scheme in Tab. 10.1):

Table 10.1 Correlation between the languages of the EPR problem and the quantum eraser scheme.

EPR spin-$\frac{1}{2}$ language	ESW quantum eraser language
Particle 2	Photon
Particle 1	Atom
Particle 2:	
Spin-up state $|\uparrow_2\rangle$	Photon in top cavity
Spin-down state $|\downarrow_2\rangle$	Photon in bottom cavity
Particle 1:	
Spin-up state $|\uparrow_1\rangle$	Atom from bottom slit
Spin-down state $|\downarrow_1\rangle$	Atom from top slit

"This is then the lesson of this review of the EPR problem: Statements such as 'particle 2 is in the spin-up state $|\uparrow_2\rangle$' already before particle 1 is found in the spin-down state $|\downarrow_2\rangle$ are just meaningless. The property of being in a pure-state spin-up particle, say, is not possessed by particle 2 at times before an up/down measurement on particle 1 has found it spin-down. This is quite analogous to the situation discussed above, where the property of being a first-slit atom is not possessed by the atom until the photon is found in the first resonator."

The interesting point is that people these days don't talk about the quantum eraser "paradox," nor should they. The problem has been well explained and understood in light of excellent experiments from various groups, as explained in Endnotes 10. To sum up, the word "paradox" in the EPR paradox is erased by the quantum eraser. It should be called the EPR–quantum eraser problem. This is further discussed in the reprints in Endnotes 10.

10.6 From Complementarity to Consciousness

"There can never be any real opposition between religion and science; for one is the complement of the other."

Max Planck

With the notions of correlation and entanglement as contained in the quantum eraser and the EPR paradox under our belt, we turn to some philosophical implications, for example, the deep question concerning possible relations between quantum mechanical complementarity, biological consciousness, and theological communion. Let us proceed by asking the quantum mechanics and philosophers to share their thoughts on these subjects with us.

First, however, allow me to make a point considering mechanics as philosophers. Some wit has said that quantum mechanics are about as interested in philosophy as auto mechanics are. Well, this diesel mechanic has spent a lifetime in the company of both groups and he agrees. Even though we "real" mechanics could choose to be offended by being lumped together with free (sometimes fuzzy) thinkers, we choose not to be.

Let us return to the arguments of Evan Walker. He now ups the ante and injects the notion of state reduction into our theosophical discussion. We have introduced the logical idea of state reduction, but never used that phrase when we discussed quantum erasure in Chapter 9. There we learned that, when we leave the shutters closed, we have no interference pattern as in Fig. 9.4. This is also the situation depicted in the middle part of the *Newsweek* article in Fig. 9.2.

As we discussed in Chapter 9, if we select one shutter to open, say the top shutter, then when the photodetector records a count, we know that the atom passed through the top hole, as in Fig. 9.5. If no count is recorded, the photon must be in the lower cavity. This is state reduction. In the quantum eraser example, it is the selection of one or the other possible states of an object (atom from top or bottom slit) as determined by observing its entangled companion in its corresponding state (photon in top or bottom cavity).

Suppose, on the other hand, that we select to open both shutters. Then, if we get a count, we know that the atom associated with an eraser is contributing to the interference pattern. That is, we have selected to know about the wave nature of the atom. To summarize: open one shutter and you have selected to receive which-path (particle) information; open both shutters and you have state selected to know which-fringe (wave) information. Walker's thoughts along these lines are summarized in the following paragraph (in our adaptation):

> "[I]n quantum mechanics, before a measurement, before observation, the state of the system is described as a collection of possibilities [e.g., which- path or which-wave possibilities]. In the case of the brain, a range of [quantum] possibilities arise – namely, in the particular synapses that have the potential to fire. [C]onsciousness is associated with the process of creating these possibilities. When observation takes place and when one state is selected, that process occurs in association with our consciousness. Finally, when the 'observation' happens – when state selection occurs – one synapse, from all those that could have fired, does fire. And the state selected by this synaptic firing, specifies just what the brain, and consequently, what the body, will do next. This observation process brings our brain's next thought and our body's next action into being. [Thus] we have found that will is quantum mechanical state selection going on in the brain."

Wow! Walker is saying that the interworkings of the brain involve quantum mechanics at some deep level. The famous Sir Roger Penrose agrees. But many quantum mechanics seem to have reservations. They think the "warm wet" brain is just too noisy and the "synaptic firing" is reduced to behaving classically. In any case, the idea of quantum mechanics going on in the brain is an interesting and open issue.

Finally, we give the stage back to the incisive Dr. Baggott. Referring to the difference between the classical and quantum pictures of the will, he says:

> "However, quantum physics may paint a rather different picture, possibly allowing us to restore some semblance of self-esteem. Out go causality and determinism, to be replaced by the indeterminism embodied in the uncertainty relations. Now the future development of a system becomes impossible to predict except in terms of probabilities. Furthermore, if we accept von Neumann's and Wigner's arguments about the role of consciousness in quantum physics, then our conscious selves become the most important things in the universe. Quite simply, without conscious observers, there would be no physical reality ..."

We are all interested in "the what and the why." It is a thrill to find clean clear statements from the deep (quantum) thinkers and a pleasure to record a few of them. Heisenberg voices a similar endorsement for the role of the observer:

> "Even the demand for objectivity, which has long ranked as the precondition for all science, has undergone restriction in atomic physics, in virtue of the fact that a complete separation of the observer from the phenomenon to be observed is no longer possible."

John Wheeler elevates this observer-observed symbiosis to a guiding principle of universal proportions. He calls it the *participatory universe*. Reality, according to Wheeler, is only created (or completed) by observation. Recall that this is in accord with the message of SEW quoted at the end of the last section. They say that the assignment of a specific note (spin up or spin down) before measurement is a "no-no". The particle does not have the spin-up or spin-down property before measurement. Quantum scientist Peter Zoeller-Greer[72] makes this point clearer:

> "Remember that the Scully[73] experiment teaches us that the past of the electron ... was created during its observation in the present. But we also understand that this reconstruction of the past leads us to more than one possibility. The past's reality 'happens' while it is being observed in the present ..."

10.7 From the Big Bang and Plato's Time to Pauli's Complementarity

10.7.1 *It All Started With a Bang*

From nothing to bang – suddenly reality is here. Why? Why is the universe here at all? No aspect of science defies our perceptions of reality and reason more than that of the big bang. Gradually, as we uncover more of nature's subtle realities, we are discovering just how closely scientific realities and subtle mysteries are related.

Steven Weinberg points to the extreme "fine tuning"[74] that had to exist before our universe could "be"; the energy of the big bang had to be controlled or tuned to one part in 10 to the power 120! Michael Turner notes that this precision:

> "... is as if one could throw a dart across the entire universe and hit a bull's-eye one millimeter in diameter on the other side."

If we knew what happened, we might truly understand how nature actually works. For it was during that brief moment that the groundwork of all to follow was laid in the "ten commandments of reality." Cosmologist Hoyle speaks.[75]

"Would you not say to yourself, 'Some supercalculating intellect must have designed the properties of the carbon atom, otherwise the chance of my finding such an atom through the blind forces of nature would be utterly miniscule.' Of course you would ..."

In like mind, Paul Davies holds forth:[76]

"I cannot believe that our existence in this universe is a mere quirk of fate, an incidental blip in the great cosmic drama ... Through conscious beings the universe has generated self-awareness. This can be no trivial detail, no minor byproduct of mindless, purposeless forces. We are truly meant to be here."

To make headway in such a situation, a rich mixture of physics and philosophy is often the best tool. It has historically proven to be the link between deep thought and proven fact. Of course, philosophy has often been misguided, but more often it has been the faint light leading us to understanding. From Pythagoras to Einstein, the best minds in science have been staunch believers in the proper utilization of philosophy as a first step to explain the seemingly unexplainable. Einstein said as much when he stated:[77]

"I hold it true, that pure thought can grasp reality, as the Ancients dreamed ... We can discover by means of purely mathematical constructions the concepts and the laws connecting them with each other, which furnish the key to the understanding of natural phenomena."

Plato would have loved Einstein's mathematical insights into the ways of the universe! If Plato were here today, he might not have all the solutions, but he would likely be sympathetic to the ideas of current physics. Measurements and models are the foundations upon which we build our picture of reality; but now in participation with (as coauthors of) that reality. Thus the thinking involved in science, as with all forms of successful thinking, should be of a straightforward, pure, but open-minded nature. Men like Einstein and Schrödinger accepted this as an inherent, self-evident truth. Life, religion, love, and math are all reality. Those who recognize this fact are those who truly embrace the gift of existence.

10.7.2 Science and Religion Were Clearly Intertwined in Early Western Philosophy

"The West got both God and Physics, the East got neither, and for the same reason."

Rothmann and Sudarshan

Science is based on faith that there is law and logic, reason and rationality, in nature. This is so despite the fact that much of recently discovered science is complex, counter- intuitive, and even seemingly mystical. The great mathematical physicist, Hermann Weyl, says:[78]

"A mathematician steps before you, speaks about metaphysics, and does not hesitate to use the name of God. That is an unusual practice nowadays. In other times, this was different. Pythagoras, whose figure almost merges into the darkness of mythology, by his fundamental doctrine that the essence of things dwells in numbers, became at the same time the head of a mathematical school and the founder of a religion. Plato's profoundest metaphysical doctrine, his doctrine of ideas, was clad in mathematical garb when he expounded it in rigorous form."

Indeed, it has been the mathematical logic of the Pythagoreans (including Plato) that has driven science through the ages. Law and order, reason and rationality, these are the ground rules of Western thought.

In times past, the Chinese thought differently. Tony Rothman and George Sudarshan[79] raise the question as to why the Chinese never invented modern science. They say:

"To the extent that the question has an answer Joseph Needham seems to opt for a theological one. Without faith in a rational creator, 'there was no conviction that rational personal beings would be able to spell out in their lesser earthly languages the divine code of laws which (the creator) had decreed aforetime.' The West got both God and Physics, the East got neither, and for the same reason."

However, modern quantum mechanics has cast things in a different light. Measurement, logic, and mathematics are still the basis for science. But the strict determinism which has guided science from the time of Aristotle until the end of the 19th century has gone by the board. Things are very different today than they were, even in the relatively recent time of Maxwell (please recall, however, that Maxwell's demon was a harbinger of modern quantum thinking).

10.7.3 *From Plato to Pauli*

> "Modern microphysics turns the observer once again into a little lord of creation in his microcosm, with the ability (at least partially) of freedom of choice and fundamentally uncontrollable effects on that which is observed.
>
> Pauli, letter to Jung (December 23, 1947)

In modern quantum mechanics, the order of the day is still experimental observation, together with mathematical explanation, as was also the case for Brahe and Kepler. Nevertheless, we are moving more toward the Asian mystic's point of view. Heisenberg says it well:[80]

> "Even the demand for objectivity, which has long ranked as the precondition for all science, has undergone restriction in atomic physics, in virtue of the fact that a complete separation of the observer from the phenomenon to be observed is no longer possible."

Thus, he concludes that we are moving away from the strict Aristotelian logic back toward Plato's way of thinking:

> "It is certainly no accident that the beginnings of modern science were associated with a turning away from Aristotle and reversion to Plato."

That is, the modern view leads us away from Aristotle back to Plato, who saw himself as a kind of "co-arranger of the universe," and whose insights were a few millennia ahead of his time. Recall that he maintained that time began with the creation of the stars. As Owen Gingerich notes, Plato thought of a "timeless eternity."

Schrödinger likewise credits Plato with the idea of an existence unbounded by time. Schrödinger asserted that subjective "statistical time" is what is most important in our lives. That is, our "mind" or "soul" exist outside of the ravages of time.[81]

Plato would have loved modern physics. He understood the need for probabilistic logic and advocated perception and ideas as part of reality. He was really on to something.

Let us pause for a moment and return to the study of Plato, the man. Plato was strongly influenced by the school of Hercleitus, who was famous for his cosmology, which held that fire was the most important and basic element of the universe. Today, we would replace the word "fire" by "hot gas" but, whatever name we tack on it, stellar matter is surely mostly "fire."

Plato's young life was not that of a quiet scholar. It is said that he was nicknamed Plato (roughly "the broad") to describe his character. Perhaps it came from the width of his shoulders, from the breadth of his thoughts, or from the size of his head. Plato began his career as a student of Socrates. He left Athens after Socrates' death and is believed to have traveled to Crete, Italy, Sicily, and even Egypt. Eventually he returned to Athens and began his own school known as the Academy. Because Plato came from a wealthy, aristocratic family, his life was spotted with involvement in political affairs, most of which he found distasteful.

Toward the end of his life, Plato became disillusioned with politics, having experienced nothing but corruption during his tenure of involvement with the state. He was not an adherent of democracy as it existed in his day in Greece, believing in the efficiency of a more centrally controlled form of government. As he said:

> "Until philosophers rule as kings or those who are now called kings and leading men genuinely and adequately philosophize, that is, until political power and philosophy entirely coincide ... cities will have no rest from evils, ... nor, I think, will the human race."
>
> Plato, *Republic* (473 c–d)

As usual, his thoughts were tempered by experience and are still regarded as deep wisdom.

Although he wanted a government career as a young man, he would decline such opportunities as he grew older. His service included four years in the military during the Peloponnesian war between Athens and Sparta. He again enlisted and fought in a later war. Plato was a real scholar and soldier, decorated for bravery in action. But he returned to a career in academia bitterly disappointed by the poor standards set by Greece's political leaders. At the Academy he hoped to inculcate the country's future political leaders with better values and higher standards. The school was structured around a diverse curriculum including mathematics, political theory, and philosophy. Its goal, in part, was to cultivate the logical and ethical thinking necessary to productively restructure the government.

Plato's main love was always philosophy, and he applied it to every aspect of life. The practices of psychology, religion, art, love, morality, government, law, and the theory of forms (i.e., ideas) all received much attention in Plato's many written works. Perhaps Plato's most significant lasting contribution is a phenomenon referred to as Platonism or (exaggerated) realism. This metaphysical philosophy addresses our perceptions of reality; the well-known example of a man facing the back wall of a cave is a case in point. Through the light of a fire, the man sees reflections from the outside of moving objects and people on the wall of the cave. The reflections are of course dull and distorted when compared to the real images outside. Plato claimed reality is like that. What we think we see, and know, is only a partial reflection of the real truth, the real objects, and the real world.

The basis of today's quantum physics can be considered to be in substantial agreement with Plato's view of reality as subjective. As we have seen in the last chapter, our *knowledge* of quantum objects affects the way we treat those objects. In other words, the state of a quantum system is determined by what we know about that system. Plato would have liked that.

Thus it is clear that Plato offers an early profoundly different "line of thought." He believed that the reality of matter, such as rocks and water, was not to be found in the matter itself. Plato held that reality reaches completion in our conscious recognition of it.[82] He asks:

> "Were it not for that recognition, would anything exist at all?"
>
> Plato

Which is an eerie precursor to the wisdom of John Wheeler, who says:

"No phenomenon is a physical phenomenon until it is an observed phenomenon."

John Wheeler

In the quantum eraser experiment we guide reality by choosing one aspect of duality. The information (spirituality) we choose to glean today decides the fate of the (material) universe tomorrow. In that regard Wolfgang Paul speaks:

"There will always be two attitudes [spiritual vs. material] dwelling in the soul of man, and the one will carry the other already within it, as the seed of its opposite ... In allowing the tension of the opposites to persist, we must also recognize that in every endeavor to know or solve we depend upon factors which are outside our control, and which religious language has always entitled 'grace'."

Paul

Indeed the notion of information as realization and observation as creation is our leitmotif. Physicist and Priest John Polkinghorne notes that it is reasonable to think that our soul, or "information file", is held eternally in God's Mind:[83)]

"It is, however, a perfectly coherent possibility to deepen the discussion by adding a theological dimension, and to affirm the belief that the God who is everlastingly faithful will preserve the soul's pattern post mortem ..."

Polkinghorne

And He may well do it using quantum mechanics. In any case, we cannot but agree with Phillips who said:

"God has given us an incredibly fascinating world to live in and to explore, [for which] I'm really thankful."

William Phillips upon winning the Nobel Prize.

Endnotes 1

$$y = 10^X$$
$$\log_{10} y = x$$

"In the moment when the right ideas emerge, an indescribable process of the highest intensity takes place in the soul of the person who sees these ideas. This is that amazed shock that Plato speaks of in the 'Phaedo' with which the soul as it were thinks back to something it has unconsciously always known."

Werner Heisenberg

Boltzmann says: $S = k \log W$.
So we need to know what logarithms are all about.
The next few pages are all you need to tote.

E1.1 Algebra: The Bare Bones

The importance and usefulness of a little basic algebra cannot be overstated. We will go slowly through the basic ideas of algebra following their development long ago. These basics are still very helpful today.

Algebra is nothing more than arithmetic plus a little logic. We begin by giving the number we seek a name, e.g., let "x" stand for any number. For example, it might be the number of miles from home, or the amount we add to our savings each month.

If we add x dollars to our account the first of every month and if we had, say, 1500 dollars last month then we will have $1500 + x$ dollars in the bank this month. If $x = 100$ dollars then we now have 1600 dollars and so on.

The famous physicist Richard Feynman told a story about his first encounter with algebra that bears repeating. His cousin was moaning about how hard algebra was. Young Richard asked, "What is algebra?" His cousin responded, "It's like this: what number plus 9 is 17?" Feynman said, "Oh, that's easy, it's 8." His cousin replied, "That's not fair you used arithmetic and you are supposed to use algebra!"

The cousin wanted to see an algebraic solution such as:

1) Let x be the unknown (a number to be determined), so that we write

$$9 + x = 17, \tag{1}$$

The Demon and the Quantum, Second Edition. Robert J. Scully and Marlan O. Scully

ISBN 978-3-527-40983-9

2) then subtract 9 from both sides of the equation

$$\underbrace{9-9}_{0}+x=\underbrace{17-9}_{8} \tag{2}$$

3) and we find $x = 8$.

The cousin was frustrated that smart Richard was not playing the game. However, if he had just upped the ante a little he would have made his point. For example, he might have said, "What number added to 9 equals 2 times that same number?" Not so easy now – but a little algebra gives the answer in a jiffy as follows:

1. The equation for x now reads

$$x+9=2x \tag{3}$$

2. Now subtract x from both sides

$$\underbrace{x-x}_{0}+9=\underbrace{2x-x}_{x} \tag{4}$$

3. So now we see $x = 9$.

Let us proceed to summarize a few of the most simple, but very useful, facts of algebra.

1. Any equation is still true if we add or subtract the same thing from both sides of the equal sign, as per the proceeding example.
2. Likewise, an equation is still true if we multiply or divide both sides of the equation by the same thing. For example, suppose we multiply both sides of the Eq. (1) by 2, then we have

$$(9+x)\times 2=17\times 2$$

So $2x + 18 = 34$ or $2x = 34 - 18 = 16$ and if we divide both sides by 2 we have

$$\frac{2x}{2}=\frac{16}{2}=8.$$

3. An expression written as a fraction (with a numerator and denominator) is the same if we multiply or divide both the numerator and denominator by the same thing. This is the same idea as canceling a common factor from the top and bottom of a fraction, for example:

$$\frac{8}{12}=\frac{4\times 2}{4\times 3}=\frac{2}{3},$$

and likewise

$$\frac{8x}{12y} = \frac{4 \times (2x)}{4 \times (3y)} = \frac{2x}{3y}.$$

Algebra is really just arithmetic and common sense, but what a powerful tool comes from that mixture!

Of course many books have been (and continue to be) written about algebra and this is far from being a math book. We will be very careful to explain any algebra we use. Rob's book is essentially algebra free. Only in these endnotes do we use a bit of it.

We say the book is "essentially" free of algebra, but Rob does present Boltzmann's famous equation relating entropy, S, to the logarithm of a certain quantity, $s = k \log w$. Therefore, we need to be friendly with logarithms and that is where we head next.

E1.2 Logarithms

A. Exponents as a Step toward Logarithms

Logarithms and exponents are closely related. We must first consider exponents before we introduce logarithms. Exponential notation is very useful. For example, simplify powers of ten as

$$100 = 10 \times 10 = 10^2 \qquad \text{(hundreds)}$$

$$1000 = 10 \times 10 \times 10 = 10^3 \qquad \text{(thousands)}$$

$$\vdots$$

$$1{,}000{,}000 = 10 \times 10 \times 10 \times 10 \times 10 \times 10 = 10^6 \qquad \text{(millions)}$$

Write any power of ten (like a thousand) as a product of 10's as 10 to some power. For example, call

$$x = 10^y.$$

Then we have

$$x = 1000 \quad \text{for } y = 3$$

and

$$x = 1{,}000{,}000 \quad \text{for } y = 6.$$

Please note also that one million equals one thousand times one thousand, which is $10^6 = 1000 \times 1000 = 10^3 \times 10^3 = 10^{3+3}$. This example illustrates the very useful fact that multiplying any two numbers, each expressed as a power of ten, is equivalent to simply *adding* the exponents.

Let us next ask what it means to consider powers of more general numbers (besides ten) like two. Well, $2 \times 2 = 4$ so we say that $2^2 = 4$ and $2 \times 2 \times 2 = 2^3 = 8$, etc. Furthermore, $8 = 2 \times 4$ and since $2 = 2^1$ and $4 = 2^2$ we have $2 \times 4 = 2^1 \times 2^2 = 2^{1+2}$. So for any number x we say $x^a \times x^b = x^{a+b}$, in words

> To multiply two numbers
> Expressed as powers of the
> Same basic number (here x),
> We just add the powers.

Next, let us consider what it means to let a and b be fractions like: $\frac{1}{2}$ or $\frac{1}{4}$ or 0.187 etc. Recall the definition of the square root $\sqrt{x}$. It is simply the number, which multiplied by itself, gives x. That is,

$$\sqrt{x} \times \sqrt{x} = x.$$

So we can write

$$x^{\frac{1}{2}} \cdot x^{\frac{1}{2}} = x^{\frac{1}{2}+\frac{1}{2}} = x.$$

Finally, we note that $\dfrac{1}{x^a} = x^{-a}$ and since $1 = \dfrac{x^a}{x^a} = x^{a-a}$ we have $x^0 = 1$.

Thus we see that a and b do not have to be simple numbers in order to be meaningful. This is a good place to introduce the concept of the logarithm.

E1.3 Logarithms in Base 10

Let us go back to the powers of 10 discussions. Recall

$$1000 = 10^3.$$

Hence we say the $\log_{10}$ of 1000 is 3, that is

$$\log_{10} 1000 = \log_{10} 10^3 = 3.$$

Any number x can be written as a power of 10, that is

$$x = 10^a, \tag{1}$$

and so we define "a" as the log of x to the base 10,

$$\log_{10} x = a. \tag{2}$$

In other words: a is the number that makes $x = 10^a$; again, if $x = 1000$ then $a = 3$.

Finally, we note that 10 is not special. We can (and do) use base two (in computers) instead of ten. Then we would write

$$x = 2^a,$$

and therefore

$$\log_2 x = a. \tag{3}$$

However, one number is special (in a way that ten and two, etc. are not). That number is "e" and it has the value

$$e = 2.718\ldots \tag{4}$$

The reason why scientists, engineers, and mathematicians like "e" so much has to do with the fact that "e" is important in describing the way things grow, decay, oscillate and much more. Hence, we hold a special place for $\log_e x$, and call it the natural logarithm. We say

$$\boxed{\begin{aligned} x &= e^a = \exp(a) \\ \log_e x &= \log x = a \end{aligned}} \tag{5}$$

That is, we drop the subscript and call $\log_e x$ just plain $\log x$.

E1.4 The Natural Logarithm

Suppose we invest our money at some annual percentage rate of interest r. Then if we invest P_0 dollars after one year we will have P_1 dollars where

$$P_1 = (1 + \alpha)P_0, \tag{6}$$

and $\alpha = \dfrac{r}{100}$, e.g., if r is 5, then $\alpha = 0.05 = 5\%$ is the fractional gain. After N years, we will have:

$$P_N = (1 + \alpha)^N P_0 \tag{7}$$

Suppose however that the interest is added into our account every $\frac{1}{2}$ year, then at the end of the year we will have

$$\mathscr{P}_1 = \left(1 + \frac{\alpha}{2}\right)^2 P_0. \tag{8}$$

Now if we add the interest n times per year we will have $\left(1+\frac{\alpha}{n}\right)^n P_0$ after one year; and after N years we would have

$$\mathscr{P}_N(\mathrm{n}) = \left(1+\frac{\alpha}{n}\right)^{nN} P_0. \tag{9}$$

Let us ask how much money we would have if we let n get very large and let $N = 1$, i.e., leave the money in for only one year. Then we would have (in an obvious notation):

$$P_1(n \to \infty) = \lim_{n\to\infty} \left(1+\frac{\alpha}{n}\right)^n P_0, \tag{10a}$$

where "$\lim_{n\to\infty}$" is the short hand for "in the limit that n is very large," and if we let $n \to \alpha m$ we have

$$P_1(m \to \infty) = \lim_{m\to\infty} \left(1+\frac{1}{m}\right)^{\alpha m} P_0. \tag{10b}$$

Hence, we define

$$e = \lim_{m\to\infty} \left(1+\frac{1}{m}\right)^m, \tag{10c}$$

so that

$$P_1(n \to \infty) = e^{\alpha} P_0. \tag{10d}$$

This is *continuous* compound interest.

The value of e as given by Eq (10b) can be estimated by taking say $n = 10$ and calculating $(1 + 0.10)^{10}$ on your calculator. This gives $e \cong 2.6$ where $\cong$ means approximately equal to. The actual value is $e = 2.718$.

E1.5* Showing that $\log(1 + x) = x$ for Small x

Thus, the value of e^x is

$$e^x = \left[\lim_{n\to\infty} \left(1+\frac{1}{n}\right)^n\right]^x \tag{11a}$$

Now let us consider

$$e^x = \lim_{m\to\infty} \left(1+\frac{x}{m}\right)^m \tag{11b}$$

* This section may be skipped on first reading.

In particular, we will have need for the fact that for small x, Eq. (11b) implies

$$e^x \cong 1 + x \quad \text{for } x \text{ small compared to 1.} \tag{12}$$

To show this we consider successive powers of m to see

$$\left(1+\frac{x}{2}\right)^2 = \left(1+\frac{x}{2}\right)\left(1+\frac{x}{2}\right) = \left(1+\frac{2x}{2}+x^2\right) \cong 1+x$$

for small x compared to 1.

$$\left(1+\frac{x}{3}\right)^3 = \left(1+\frac{x}{3}\right)\left(1+\frac{x}{3}\right)\left(1+\frac{x}{3}\right)$$

$$= 1 + \frac{3x}{3} + \text{higher order terms in } x^2 \text{ and } x^3 \text{ which are small}$$

for small x

$$\cong 1 + x$$

$$\vdots$$

$$\left(1+\frac{x}{m}\right)^m = 1 + m\frac{x}{m} + \underbrace{\cdots}_{\text{Small stuff}}$$

$$\cong 1 + x$$

Taking the logarithm of both sides of Eq. (12) yields

$$\log e^x = \log(1+x), \quad \text{for small } x$$

and since $\log e^x = x$ we have

$$\boxed{x = \log(1+x)} \quad \text{for } x \text{ small.} \tag{13}$$

Key Points

- The natural number $e = 2.718\ldots$ is defined by the equation

$$e = \lim_{n\to\infty}\left(1+\frac{1}{n}\right)^n$$

- We use the symbol log to mean $\log_e$. Most books call $\log_e x = \ln x$.
- For small values of x, we have derived the useful relation:

$$x = \log(1+x)$$

Endnotes 2

Momentum: $p = mv$

Energy: $E = \frac{1}{2}mv^2$

Temperature: $\frac{1}{2}kT = \frac{1}{2}mv^2_{\text{average}}$

"Beim Sirok der Sonnenwagen
Purpurrot sich niedersenkt:
Da gebt der Natur die Ehre,
Froh, an Aug' und Herz gesund."

Goethe

"When at dusk the Sun is driven
down in crimson fire glow
There in Nature's deepest kernel
healthy, glad of heart and sight
you perceive the great eternal
essence of chromatic light."

Translation by Victor Weisskopf and Dennis Worth

E2.1 Momentum and Energy

Momentum is defined as mass m of an object (atom, auto etc.) times its velocity v. That is, the momentum p is given (in one dimension) by:

$$p = mv \quad \text{(one dimension)}. \tag{1}$$

This is to be compared with an object's kinetic energy, E, which is given by

$$E = \frac{1}{2}mv^2. \tag{2}$$

Both momentum and kinetic energy are closely involved with the forces acting on the object. For example, Newton's second law tells us that the force, F, on an

The Demon and the Quantum, Second Edition. Robert J. Scully and Marlan O. Scully

ISBN 978-3-527-40983-9

object will produce an acceleration, a, defined by the change in velocity $\Delta v - v_2 - v_1$ divided by the change in time $\Delta t = t_2 - t_1$. In general, we mean by Δx the change in x between two points 1 and 2.

$$F = ma = m\frac{\Delta v}{\Delta t} \tag{3}$$

where Δv is the change in velocity occurring in time Δt, i.e. $\Delta v = v(\Delta t) - v(0)$. Please note that $m\Delta v$ is just the change in momentum $\Delta p = m\Delta v = m[v(\Delta t) - v(0)]$ so re-write Newton's law as

$$F = \frac{\Delta p}{\Delta t} \quad \text{(one dimension)}. \tag{4}$$

The force acting through a distance Δx defines the work done on an object; that is, $W = F\Delta x$. If no other forces act upon the particle, e.g., no gravitational or frictional forces, then the work done will equal the kinetic energy. In such a case we may write

$$E = W = F\Delta x. \tag{5}$$

Now suppose we boost the particle with a constant force from zero velocity to velocity v in a short time Δt. The distance moved, Δx, will be the average velocity, $\frac{v}{2}$, times the boost time Δt, that is

$$\Delta x = \frac{v}{2}\Delta t. \tag{6}$$

If the velocity is initially zero, then, the force F is related to the velocity $v = \Delta v$ by Eq. (3) so that $F = \frac{mv}{\Delta t}$, hence the kinetic energy is given by

$$E = F\Delta x = \left(\frac{mv}{\Delta t}\right)\left(\frac{v}{2}\Delta t\right) = \frac{1}{2}mv^2. \tag{7}$$

E2.2 Connections between Kinetic Energy and Temperature

Suppose we have a gas of atoms that are constrained to move in one direction such as in Fig. 2E-1.

The average kinetic energy per atom is governed by the absolute temperature, T, of the gas times Boltzmann's constant, k, which is

$$\frac{1}{2}(mv^2)_{\text{average}} = \frac{1}{2}kT \quad \text{(one dimension)}. \tag{8}$$

The constant k is important because it relates the mechanically constructed "energy" to the thermodynamic concept of temperature. It has the numerical value of $k = 1.38 \times 10^{-23}$ Joules/Degree absolute. By "absolute" we mean that at $T = 0$ there will be no thermal motion. This temperature is $-273°$ Centigrade.

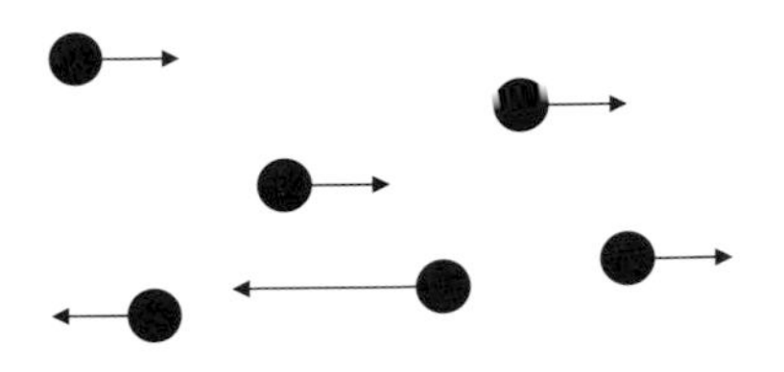

Fig. 2E-1 An ideal gas of non-interacting particles is moving only in the horizontal direction. Each atom has some velocity, v_1 for the first particle, v_2 for the second particle and so on.

E2.3 Trigonometry

"Let no one ignorant of geometry enter here."

Greek Temple Inscription

Just as we introduced a little algebra in Chapter 1, as a tool for latter chapters, so we now present the very bare bones of trigonometry, which will be likewise useful. In the next section, we introduce the basic ideas of trigonometry as they arose in the surveying of land. Then we follow the Greeks who used simple trig to measure the radius of the earth (yes, they knew the earth was round!). Finally, we examine a more involved application that illustrates how to use simple trigonometry to measure the distance to the stars. Vectors will be discussed, using a bit of trig, in Section E2.4 and also in the end notes to Chapter 8.

A. Introduction

The use of triangles in geometry and triangulation in technology is a good example of the efficiency of mathematics. The ancients learned early to fix a point by measurements made far from the point. For example, suppose we want to fix a boundary point of our land after the Nile has flooded and washed away all markers. Fig. 2E-2 illustrates one method for doing this. There we see that by measuring two angles and one length (out of the flood plain) we can easily fix a boundary point. This simple procedure is possible because we have the nice triangle in which two sides of equal length, d, meet at 90°. A triangle having one side of 90° is called a right triangle.

However, the usual case is that we are not so lucky as to have a straight shot due south of our boundary corner for a distance d. Maybe there is a river in the way. Therefore, we have to think a little bit more, as in Fig 2E-3.

In general, we call the ratio D to d in Fig. 2E-3 tan A, that is

$$\tan A = \frac{D}{d}$$

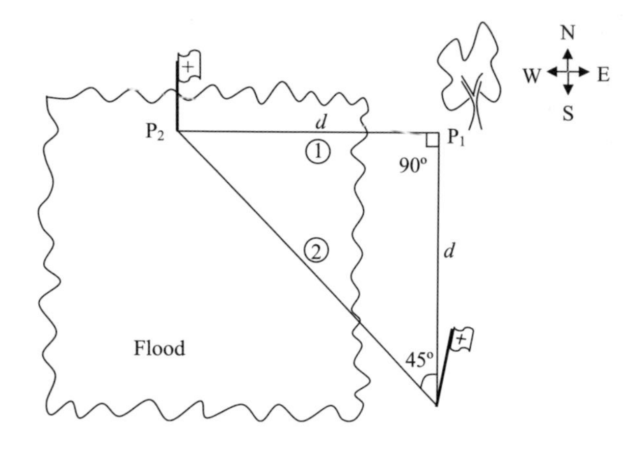

Fig. 2E-2 By starting at the point P_1, e.g., a tree (not affected by the flood) and shooting a 90° angle due west, then walking a distance *d* due South and sighting at 45° to N-S line, the point P_2 is where the two lines marked ① and ② cross will be fixed.

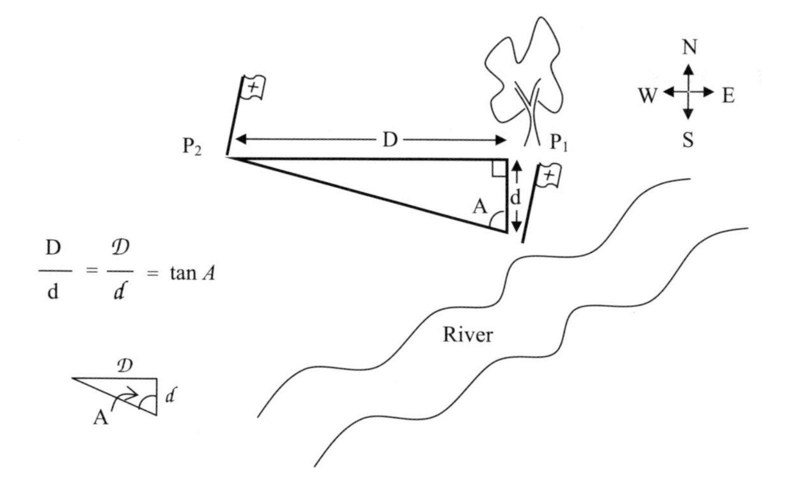

Fig. 2E-3 What if we can't use the method of Fig. 2E-2? Now we measure the distance *d* and the angle *A*. The method of equivalent triangles determines the distance *D*. That is, as depicted in the figure, since the angles *A* in the two triangles are equal, we can find *D* once we know *d*, given $\mathcal{D}$ and $\mathcal{d}$.

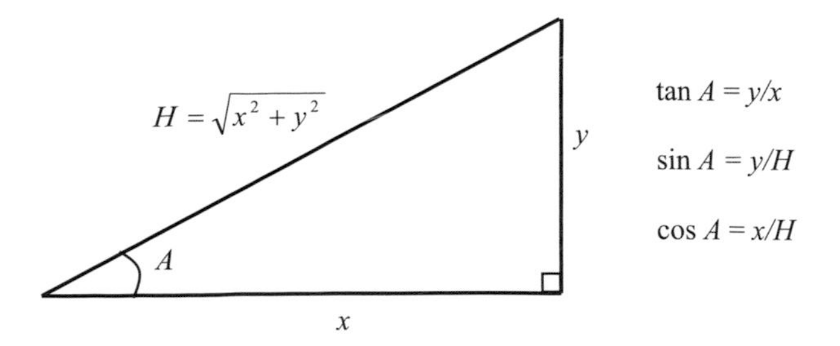

Fig. 2E-4 The three trig relations. The sin A and cos A are defined as the ratio of the side opposite the hypotenuse, H, and the side adjacent to the hypotenuse respectively. The Pythagorean theorem, $H^2 = x^2 + y^2$, is indicated.

or

$$D = d \tan A.$$

The other two most useful trig functions are the sine and cosine of an angle as defined in Fig. 2E-4.

E2.4 Vectors

The velocity of a particle is a vector; it has direction and magnitude. Its mass is a scalar, therefore described by one number. Consider the fact that a particle moving in the x direction is clearly doing something different from one moving in the y direction. In general, a particle moving at some angle θ to the x-axis is described by a vector $\vec{v}$, given by

$$\begin{aligned} \vec{v} &= \hat{x}v_x + \hat{y}v_y \\ &= \hat{x}v \cos\theta + \hat{y}v \sin\theta, \end{aligned} \tag{9}$$

where $\hat{x}$ and $\hat{y}$ are unit vectors (have unit length) and point along the x and y directions; see Fig. 2E-5.

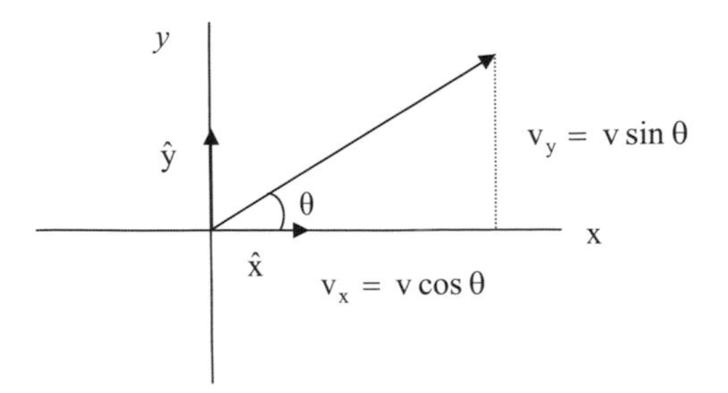

Fig. 2E-5 A two-dimensional vector $\vec{v}$.

More generally, write a vector in three dimensions as

$$\vec{v} = \hat{x}v_x + \hat{y}v_y + \hat{z}v_z \tag{10}$$

in an obvious extension of the 2-D vector of Eq. (9).

Momentum, like velocity, is clearly a vector given by

$$\vec{p} = m\vec{v}, \tag{11}$$

however, energy is not a vector, and kinetic energy is given by

$$E = \frac{1}{2}m(v_x^2 + v_y^2 + v_z^2). \tag{12}$$

Finally, the same arguments that lead to Eq. (2.8) (the energy-temperature relation) in one dimension, generalize in 3-D to

$$E_{\text{thermal}} = \frac{3}{2}kT. \quad \text{(3 dimensions)} \tag{13}$$

Key Points

The important (1D) relations given at the beginning of this endnote read in 3D:

- Momentum: $\vec{p} = m\vec{v}$
- Energy: $E = \frac{1}{2}m(v_x^2 + v_y^2 + v_z^2)$.
- Temperature: $\frac{3}{2}kT = E_{\text{average}}$

Endnotes 3

$$e_{\text{Carnot}} = 1 - \frac{T_c}{T_h}$$

$$S_{\text{Clausius}} = \frac{\Delta H}{T}$$

"[Thermodynamics] is the only physical theory of a general nature of which I am convinced that it will never be overthrown."

Albert Einstein

The beginning of thermodynamics as a science is traced to Carnot and the second law. His ideal engine (1824) converted heat from a hot energy source at temperature T_h while rejecting waste heat to a colder bath (entropy sink) at temperature T_c. The first law of thermodynamics came later (1843–1848) at the hands of Joule. Clausius made the important step of realizing that the heat taken in on the isothermal power stroke ΔH divided by the temperature T was a fundamental physical quantity he called "entropy." Einstein's miraculous 1905 year was overwhelmingly driven by his interest in entropy and the physics of thermodynamical fluctuations.

E3.1 Reference Material on Carnot

The material on Carnot, born 1796, derives from his book *Reflections on the Motive Power of Heat and on Machines Fitted to Develop This Power.*[1] Another source we have enjoyed is H. Bent, *The Second Law*, Oxford Press, 1965. The following paragraph gleaned by Bent from the writings of Sadi's brother Hyppolyte provides an excellent glimpse into the character of our hero Capt. Carnot. Carnot, while of delicate constitution, managed to increase his strength through a strict physical regime, including a wide variety of stimulations from fencing to gardening, and dancing to violin.

"On one occasion, his brother, Hyppolyte, wrote Carnot was out walking when a horseman 'who was evidently intoxicated, passed along the street at a gallop, brandishing his saber and striking down the passer-by. Sadi darted forward, cleverly avoided the weapon of the soldier, seized him by the leg, threw him to the earth and

The Demon and the Quantum, Second Edition. Robert J. Scully and Marlan O. Scully

ISBN 978-3-527-40983-9

laid him in the gutter, then continued on his way to escape from the cheers of the crowd, amazed at this daring deed.' Not long thereafter Sadi Carnot died of cholera following an attack of scarlet fever, at the age of thirty-six."

E3.2 Force on Piston Due to One Hot Atom

When a small object, like an atom, collides with a massive piston it bounces back (reverses direction) without changing its speed. To see this, think of a marble dropped on a smooth cement surface. The marble bounces back to essentially the same height from which it began. Thus, as per Fig. 3E-1, the change in momentum is $2mv$ for an atom of mass m and velocity v reflecting off the piston.

Now in a time Δt the atom will travel a distance $v\Delta t$. The roundtrip distance is $2L$, see Fig. 3E-1, and therefore there will be $N = v\Delta t/2L$ collisions in a time Δt. This means there will be a total force on the piston of

$$F_{\text{piston}} = [\ \underbrace{(2mv)}_{\substack{\text{Momentum}\\ \text{change per}\\ \text{collision}}} \times \underbrace{v\Delta t/2L}_{\substack{\text{Number of}\\ \text{Collisions}}}\ ]/\Delta t, \tag{1}$$

where we have used $F = \Delta p/\Delta t$.

Canceling common terms in the numerator and denominator, we have:

$$F_{\text{piston}} = \frac{mv^2}{L}. \tag{2}$$

Finally we replace the mv^2 term by its average value kT, as per Eq. (8) in the endnotes of Chapter 2, to obtain the average force:

$$\boxed{F_{\text{piston}} = \frac{kT}{L}} \tag{3}$$

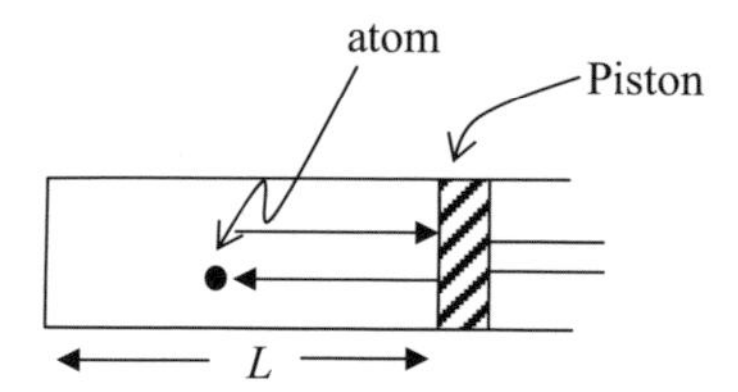

Fig. 3E-1 The initial momentum mv is reflected off the piston with final momentum $-mv$; so that the change in momentum per collision is $\Delta p = mv - (-mv) = 2mv$.

E3.3 Useful Work from Carnot Cycle Engine

Let us first calculate the work done on the isothermal expansion (power) stroke of Fig. 2b in Chapter 3. Note that the force on the piston decreases as L increases with temperature T held constant. However, if we require the expansion, ℓ (on the $1 \rightarrow 2$ part of the cycle of Fig. 3E-2 at the end of the endnotes III) to be small compared to L, we can take the force to be constant. Then the work done by the atom hitting the piston as the piston moves a distance ℓ is just

$$W(1 \rightarrow 2) = F\ell = \frac{kT}{L}\ell, \tag{4}$$

where we have used Eq. (3) for F.

Now the heat energy taken in to produce this work is, by the conservation of energy, equal to $W(1 \rightarrow 2)$ and so we may write

$$H_{\text{in}} = \frac{k}{L} T_{\text{hot}}\ell. \tag{5}$$

We have added the notation T_{hot} to remind us that the expansion stroke takes energy from the "hot" energy source.

Next, we break contact with the hot energy source and allow the atom to transfer some of its kinetic energy to the piston. In particular, we continue the expansion until the temperature of the atom reduces to that of the lower temperature "cooling water" T_{cold}. The expansion is carried out in an extremely slow (reversible) way.

The next step ($3 \rightarrow 4$ of Fig. 3E-2) is isothermal compression. The result is similar to that obtained in the ($1 \rightarrow 2$) expansion, but now we are doing compression work *on* the atom. Thus, the atom is being heated (by the work Rob does on the atom) and we remove this heat by being in good thermal contact with the coolant that is at temperature T_{cold}. The work done on the atom is then equal to the heat rejected to the coolant, that is

$$H_{\text{out}} = W(3 \rightarrow 4) = \frac{kT_{\text{cold}}}{L}\ell. \tag{6}$$

Finally, in the $4 \rightarrow 1$ part of the cycle, we complete the cycle by breaking contact with the coolant and compressing the single atom gas, thereby raising its temperature from T_{cold} to T_{hot}.

We are now in a position to calculate the efficiency of our single atom Carnot engine. Calling the efficiency, e, and recalling the definition of efficiency from the Chapter 3, we have

$$e = \frac{H_{\text{in}} - H_{\text{out}}}{H_{\text{in}}}. \tag{7a}$$

Taking H_{in} and H_{out} from Eq's (5) and (6), we arrive at

$$e = \left[\frac{kT_{\text{hot}}}{L}\ell - \frac{kT_{\text{cold}}}{L}\ell\right] \Big/ \frac{kT_{\text{hot}}}{L}\ell, \tag{7b}$$

or

$$\boxed{e = 1 - \frac{T_{\text{cold}}}{T_{\text{hot}}}} \tag{7b}$$

This is Carnot's famous result, here determined by our single atom engine. Using many atoms, i.e., a many atom gas, changes nothing; the efficiency is still given by (7b).

E3.4 From Carnot's Engine Comes Clausius' Entropy

In the previous note, we showed that the efficiency of an ideal Carnot cycle engine has the two equivalent forms (7a) and (7b). Thus, we may write

$$e = \frac{H_{\text{in}} - H_{\text{out}}}{H_{\text{in}}} = \frac{T_{\text{hot}} - T_{\text{cold}}}{T_{\text{hot}}} \tag{8}$$

and multiplying both sides by $H_{\text{in}} T_{\text{hot}}$ we have

$$T_{\text{hot}}(H_{\text{in}} - H_{\text{out}}) = H_{\text{in}}(T_{\text{hot}} - T_{\text{cold}}). \tag{9}$$

Subtracting $T_{\text{hot}} H_{\text{in}}$ from both sides and dividing by $T_{\text{hot}} \times T_{\text{cold}}$ we have

$$\boxed{\frac{H_{\text{out}}}{T_{\text{cold}}} = \frac{H_{\text{in}}}{T_{\text{hot}}}} \tag{10}$$

Clausius called this ratio the change in "entropy" and denoted entropy by "S". For this ideal reversible cycle, the total entropy change is zero, that is

$$\text{Power Stroke} \quad \Delta S(1 \to 2) = \frac{H_{\text{in}}}{T_{\text{hot}}}, \quad \left[\begin{array}{c}\text{Entropy } \textit{increases} \text{ because} \\ H_{\text{in}} \text{ is a positive number.}\end{array}\right] \tag{11a}$$

$$\text{Compression Stroke} \quad \Delta S(3 \to 4) = \frac{H_{\text{out}}}{T_{\text{cold}}}. \quad \left[\begin{array}{c}\text{Entropy } \textit{decreases} \text{ because} \\ H_{\text{out}} \text{ is a negative number.}\end{array}\right] \tag{11b}$$

E3.5 From One Atom to N Atoms: The Ideal Gas Law

Every field has its standard examples, and trial cases. For the biologist it's the fruit fly, for the thermodynamicist it is the ideal gas. An ideal gas is one whose atoms or molecules hit (interact with) the walls but do not hit (interact with) each other. We can easily understand that the force exerted on a piston by N atoms from Eq. (3) is just N times as large as that from one atom. That is, the force due to N atoms is

$$F(N \text{ atoms}) = NF(\text{one atom}) = \frac{NkT}{L} \tag{12}$$

Define the pressure as the force per unit area

$$P = \frac{F}{A}, \tag{13}$$

and note that the volume is given by $V = AL$, so dividing both sides of Eq. (12) by A and using the definitions of P and V yields the celebrated ideal gas law

$$\boxed{P = \frac{NkT}{V}} \tag{14}$$

Naturally, Eq. (14) is only an approximation to real gases because atoms do interact with each other. Nevertheless, it is a very useful result. Furthermore, we have cheated a little in "deriving" Eq. (14) – atoms live in three dimensions not one. However, the result is still true for three dimensions. This result is independent of the area A. We have taken the reader's valuable time to obtain (14) since it is so important and used extensively in thermodynamics and other fields, e.g. chemistry.

E3.6 Entropy Change in Isothermal Expansion

We conclude with a simple result obtained from the ideal gas law Eq. (14) which, we will prove in E4.2. There we will show that the change in the entropy of an ideal gas at constant temperature when we expand from a smaller volume, v, to a larger one, V, is

$$\Delta S \text{ (isothermal expansion)} = k \log\left(\frac{V}{v}\right). \tag{15}$$

Suppose, for example, $V = 2v$; then we have the important special case

$$\Delta S \text{ (isothermal expansion)} = k \log 2, \tag{16}$$

which we will meet again later.

A final note: for arbitrary expansion, $v \rightarrow V$, we define $V/v = W$, and write

$$\boxed{\Delta S = k \log W} \tag{17}$$

A glimpse of great things to come!

Key Points

- $S = k \log W$ (is written on Boltzmann's tombstone).
- $W_{12} = V_{final}/V_{initial}$ (isothermal power stroke)
- $W_{34} = V_{initial}/V_{final}$ (isothermal compression stroke)
- $\Delta S_{compression} = -\Delta S_{expansion}$
- The following diagram for the Carnot cycle engine says it all:

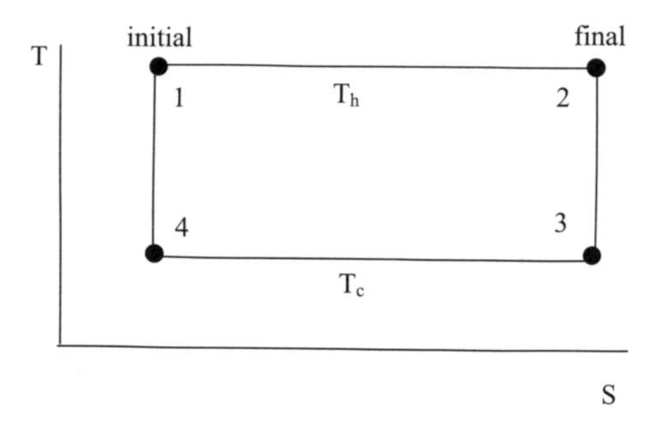

Fig. 3E-2

Endnotes 4

$$\Delta S = k \log W$$

Planck wrote this first, not Boltzmann.

> "The logarithmic connection between entropy and probability was first stated by L. Boltzmann in his kinetic theory of gases. Nevertheless, [*my equation given above*], differs in its meaning from the corresponding one of Boltzmann in two essential points.
>
> Firstly, Boltzmann's equation lacks the factor *k*. [Secondly] and this is of greater consequence, Boltzmann leaves an additive constant undetermined in the entropy *S* ... In contrast with this we assign a definite absolute value to the entropy *S*. This is a step of fundamental importance, which can be justified only by its consequences ... This step leads necessarily to the 'hypothesis of quanta'..."
>
> Max Planck

E4.1 Entropy Quantified

As advertised in the text and in E3.6, we here derive the key entropy equation

$$\Delta S = S_{\mathrm{f}} - S_{\mathrm{i}} = k \log \frac{V_{\mathrm{f}}}{V_{\mathrm{i}}},$$

where $S_{\mathrm{f}}(S_{\mathrm{i}})$ is the entropy of the final (initial) state of the gas and $V_{\mathrm{f}}(V_{\mathrm{i}})$ is the final (initial) volume. The usual derivation uses simple calculus but we will avoid following that route here. Instead, we will add up all the small changes of entropy coming from the piston moving a small distance Δx, as depicted in Fig. 4E-1.

From E3.4, we learn the so-called "entropy" generated on each step is $\Delta S = \Delta H/T$. Here ΔH is the energy taken from the heat bath in order to maintain the average kinetic energy of the atom at $\frac{1}{2}mv^2 = \frac{1}{2}kT$ as it does work on the piston. We found earlier that the work done by the atom, and therefore the energy taken from the heat bath, is given by $\Delta H = kT\frac{\Delta x}{x}$ in expanding from x to $x + \Delta x$. The entropy change on each step becomes $\Delta S = \left[kT\frac{\Delta x}{x}\right]\frac{1}{T} = k\frac{\Delta x}{x}$.

The Demon and the Quantum, Second Edition. Robert J. Scully and Marlan O. Scully

ISBN 978-3-527-40983-9

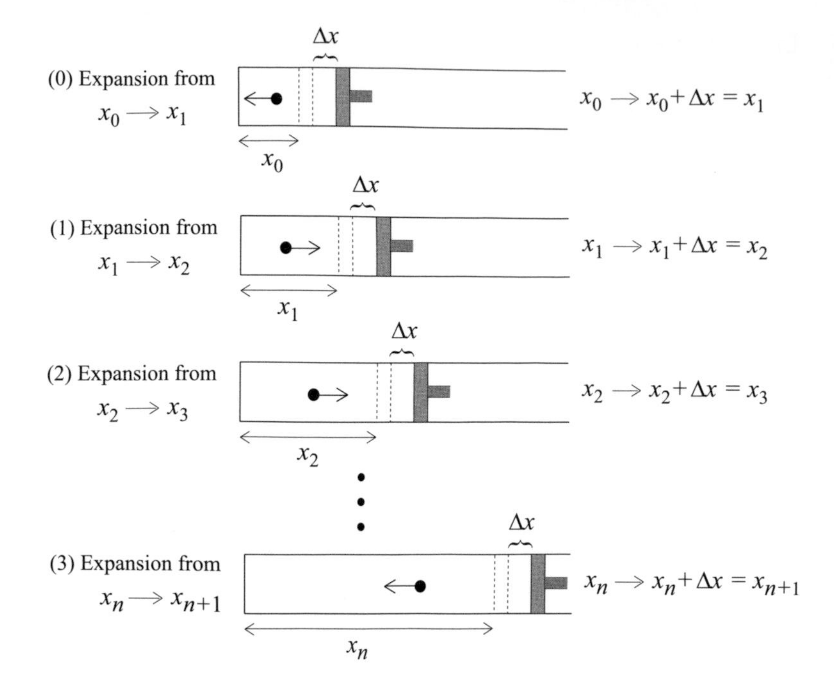

Fig. 4E-1 Single atom Carnot engine depicted at several stages as the piston moves a (small) distance, Δx, from step to step. It is easy to write the entropy generated at each step as discussed in the text.

We now proceed to write the entropy generated at each stage of expansion as follows:

$$(0)\quad \Delta S_{10} = \frac{\Delta H_{10}}{T} = \left[kT\frac{\Delta x}{x_0}\right]\frac{1}{T} = k\frac{\Delta x}{x_0}, \tag{1}$$

$$(1)\quad \Delta S_{21} = \frac{\Delta H_{21}}{T} = \left[kT\frac{\Delta x}{x_1}\right]\frac{1}{T} = k\frac{\Delta x}{x_1}, \tag{2}$$

$$(2)\quad \Delta S_{32} = \frac{\Delta H_{32}}{T} = \left[kT\frac{\Delta x}{x_2}\right]\frac{1}{T} = k\frac{\Delta x}{x_2}, \tag{3}$$

$$\vdots$$

Now we are interested in the total entropy change in expanding from the initial length $x_{\mathrm{i}} = x_{\mathrm{initial}} = x_0$ to the final $x_{\mathrm{f}=x_{\mathrm{final}}}$, this is given by

$$\Delta S_{\mathrm{fi}} = \Delta S_{10} + \Delta S_{21} + \Delta S_{32} + \cdots + \Delta S_{n,\,n+1}. \tag{4}$$

where the three dots (...) means "and so on." Then we use the result from E1.5:

$$\varepsilon = \log(1+\varepsilon) \quad \text{for small } \varepsilon, \tag{5}$$

to write Eqs. (1) to (3) as follows:

$$\Delta S_{10} = k\frac{\Delta x}{x_0} = k \log\left(1 + \frac{\Delta x}{x_0}\right) = k \log\left(\frac{x_0 + \Delta x}{x_0}\right) \tag{6}$$

$$\Delta S_{21} = k\frac{\Delta x}{x_1} = k \log\left(1 + \frac{\Delta x}{x_1}\right) = k \log\left(\frac{x_1 + \Delta x}{x_1}\right) \tag{7}$$

$$\Delta S_{32} = k\frac{\Delta x}{x_2} = k \log\left(1 + \frac{\Delta x}{x_2}\right) = k \log\left(\frac{x_2 + \Delta x}{x_2}\right) \tag{8}$$

$$\vdots$$

Next we add up the entropy changes given by Eqs. (6) to (8) etc., and use the fact that $x_1 = x_0 + \Delta x$, $x_2 = x_1 + \Delta x$, etc., together with the property of logarithms that (see E.1.2)

$$\log \frac{A}{B} = \log A - \log B,$$

to write

$$\Delta S_{\mathrm{fi}} = k \log\left\{\left(\frac{\cancel{x_0 + \Delta x}}{x_0}\right)\left(\frac{\cancel{x_1 + \Delta x}}{\cancel{x_1}}\right)\left(\frac{x_2 + \Delta x}{\cancel{x_2}}\right)\cdots\right\}$$

$$\text{since } x_1 = x_0 + \Delta x \text{ and } x_2 = x_1 + \Delta x \text{ etc.} \tag{9}$$

In this way we find

$$\Delta S_{\mathrm{fi}} = k \log\left(\frac{x_{\mathrm{f}}}{x_{\mathrm{i}}}\right),$$

where we have called the initial distance x_{i} and the final distance x_{f}.

Finally we multiply the top and bottom of the ratio $\frac{x_{\mathrm{f}}}{x_{\mathrm{i}}}$ by the piston area A to write

$$\frac{x_{\mathrm{f}}}{x_{\mathrm{i}}} = \frac{V_{\mathrm{f}}}{V_{\mathrm{i}}} \tag{10}$$

and Eq. (9) becomes

$$\Delta S_{\mathrm{fi}} = k \log \frac{V_{\mathrm{f}}}{V_{\mathrm{i}}}, \tag{11}$$

or $S_{\mathrm{final}} - S_{\mathrm{initial}} = k \log V_{\mathrm{final}} - k \log V_{\mathrm{initial}}$.

Key Point

- The Planck-Boltzmann entropy is

  $$S = k \log W.$$

Endnotes 5

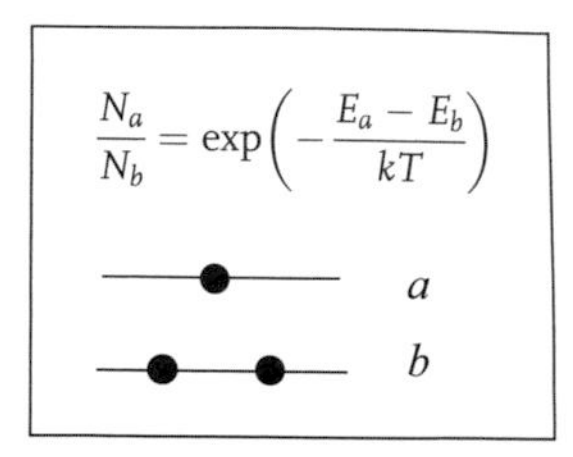

"One scientific epoch ended and another began with James Clerk Maxwell."

Albert Einstein

"From a long view of the history of mankind the most significant event of the nineteenth century will be judged as Maxwell's discovery of the laws of electrodynamics."

Richard Feynman

In this book it is Maxwell's thermodynamics that concerns us most. But it is Maxwell's work in electricity and magnetism (electrodynamics) that is most famous.

E5.1 The Great Thing We Learn from Maxwell and Boltzmann

To say *the* great thing that we learn from Maxwell *or* Boltzmann would be silly because there are *many* things we learn from both of them. There is, however, one great thing associated with Maxwell *and* Boltzmann. That is the answer to the question: what is the average number of particles having energy, E_a, in a gas of N particles at temperature T? The answer is the famous Maxwell-Boltzmann (MB) distribution.

One simple way to motivate the Maxwell-Boltzmann distribution is to start with the Planck-Boltzmann's entropy expression given by Eq. (5) and use the entropy relation from Eqs. (11a, b) of reaction E3.4

$$\Delta S = \frac{\Delta H}{T}, \tag{1}$$

Where ΔS is the change in entropy of the system (e.g. our atom in the Carnot engine) produced by adding an amount of heat energy ΔH at temperature T. Let us

The Demon and the Quantum, Second Edition. Robert J. Scully and Marlan O. Scully

ISBN 978-3-527-40983-9

write the energy as $\Delta H = \Delta E$. Now the change in entropy $\Delta S = S_2 - S_1$ is clearly given by Eq. (5) as

$$\Delta S = S_2 - S_1 = k(\log W_2 - \log W_1) = k \log \frac{W_2}{W_1}, \tag{2}$$

Then Eqs. (1) and (2) give us

$$\Delta S = \frac{\Delta E}{T} = k \log \frac{W_1}{W_2} = -k \log \frac{W_2}{W_1} \tag{3}$$

or

$$\frac{W_2}{W_1} = \exp\left(-\frac{\Delta E}{kT}\right). \tag{4}$$

Thus, when we add an energy $\mathbf{\Delta E}$, the ratio $\frac{W_2}{W_1}$ equals the ratio of probabilities

$$\frac{P(E_a)}{P(E_b)} = \frac{\exp\left(-\frac{E_a}{kT}\right)}{\exp\left(-\frac{E_b}{kT}\right)} = \exp\left(-\frac{E_a - E_b}{kT}\right) \tag{5}$$

So that if we call $E_b = E_o$ and $E_a = E_o + \Delta E$ then

$$\frac{P(E_o + \Delta E)}{P(E_o)} = \exp\left(-\frac{\Delta E}{kT}\right) \tag{6}$$

which is in accordance with the MB distribution expression Eq. (4) of the previous section.

A particularly important result concerns the ratio probabilities that an atom will be in the upper of two states having higher energy E_a and lower energy E_b, which is given by

$$\frac{P(E_a)}{P(E_b)} = \exp\left(-\frac{E_a - E_b}{kT}\right). \tag{7}$$

Key Points

- For any two energy states

$$\frac{P(E_a)}{P(E_b)} = \exp(-(E_a - E_b)/k).$$

Endnotes 6

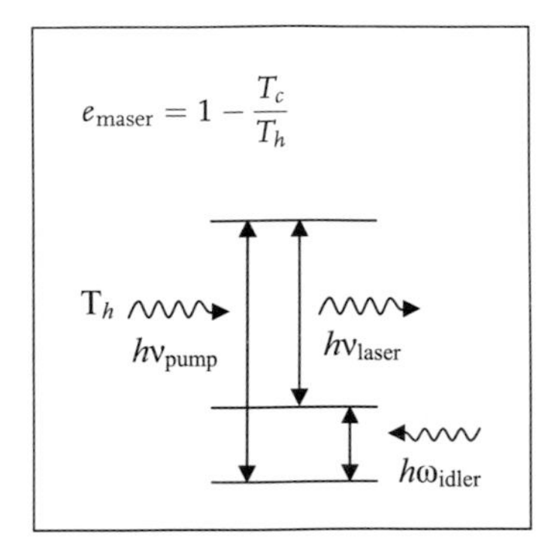

"During [the process of preparing an article on negative temperatures] I was shocked to find I was writing a paper contrary to one of the most popular statements of the second law of thermodynamics. My paper is consistent with most statements of the second law, like the principle of increasing entropy, but it is contrary to the Kelvin-Planck statement that it is impossible to operate an engine in a closed cycle that will do nothing other than extract heat from a reservoir with the performance of an equivalent amount of work."

Norman Ramsey

E6.1 The Maser/Laser as a Quantum Heat Engine

Lasers and masers operate by putting more atoms or molecules in the upper level than in the lower. When this happens, more photons are generated by stimulated emission (from the excited atoms) than are absorbed by atoms in the lower level; this is depicted in Fig. 6E-1. Thus when we have the number of atoms in the upper level n_a equal to the number in the lower level, n_b, we are at the threshold (break even) point.

Consider the situation sketched in Fig. 6E-2 where we promote the atoms from the ground state, g, to state a by putting the atoms in contact with a hot bath, at temperature T_h, but the bath does not interact with the atoms in b. We can do this in various ways, e.g., by using flash lamps.

Likewise, we couple the ground state to the b state via a cold bath at Temperature T_c. We may think of these baths as consisting of atoms having two levels as in

The Demon and the Quantum, Second Edition. Robert J. Scully and Marlan O. Scully

ISBN 978-3-527-40983-9

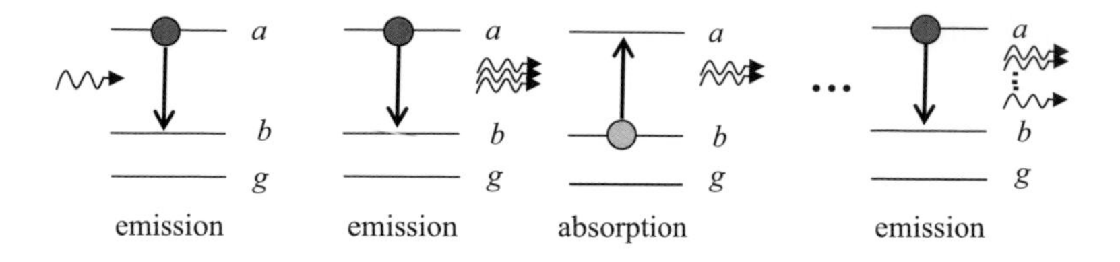

Fig. 6E-1 Figure showing that when there are more molecules or atoms in the excited state a than in the lower state b, the atoms in the laser will emit more light than they absorb.

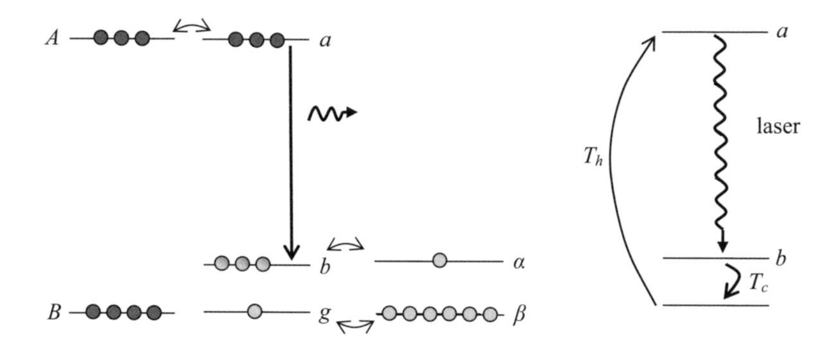

Fig. 6E-2 Hot two-level atoms having levels A and B drive population from $g \rightarrow a$ and the cold bath atoms having levels α and β which tend to pull atoms out of b and put them in g.

Fig. 6E-2. The number of atoms in a and b then obey rate equations (using upper case (A, B) and Greek (α, β) letters to denote bath atom populations and lower case letters (n_a, n_b) to denote working atom populations):

$$\left.\begin{array}{l}\text{Rate of increasing population}\\ \text{in level } a \text{ due to hot bath}\end{array}\right\} = \mathscr{H}[N_A n_a - N_B n_g], \tag{1}$$

$$\left.\begin{array}{l}\text{Rate of increasing population}\\ \text{in level } b \text{ due to cold bath}\end{array}\right\} = \mathscr{C}[N_\alpha n_b - N_\beta n_g], \tag{2}$$

where $\mathscr{H}$ and $\mathscr{C}$ are (uninteresting) rates of promotion and demotion which are governed by the heating and cooling interactions with the hot and cold baths. At equilibrium, the ratio given by Eqs. (1) and (2) equal to zero and Eq. (1) yields

$$\frac{n_a}{n_g} = \frac{N_A}{N_B}, \tag{3}$$

and Eq. (2) gives us

$$\frac{n_b}{n_g} = \frac{N_\alpha}{N_\beta}. \tag{4}$$

Now we write the key ratios (3) and (4) as determined by the hot and cold bath temperatures as

$$\frac{n_a}{n_g} = \frac{N_A}{N_B} = \exp(-\varepsilon_a/kT_h) \tag{5}$$

and

$$\frac{n_b}{n_g} = \frac{N_\alpha}{N_\beta} = \exp(-\varepsilon_b/kT_c). \tag{6}$$

Hence at threshold, where $n_a = n_b$, we have

$$1 = \frac{n_a}{n_b} = \frac{n_a}{n_g} \cdot \frac{n_g}{n_b} = \exp\left[-\frac{\varepsilon_a}{kT_h} + \frac{\varepsilon_b}{kT_c}\right], \tag{7}$$

so

$$\frac{\varepsilon_a}{kT_h} - \frac{\varepsilon_b}{kT_c} = 0 \tag{8}$$

or

$$\frac{\varepsilon_b}{\varepsilon_a} = \frac{T_c}{T_h}. \tag{9}$$

For every photon (having energy $\varepsilon_a - \varepsilon_b$) emitted, we must absorb one quantum of energy ε_a from a hot bath atom, thus we have an efficiency

$$e_{\text{maser}} = \frac{\varepsilon_a - \varepsilon_b}{\varepsilon_a} = 1 - \frac{\varepsilon_b}{\varepsilon_a}, \tag{10}$$

and using Eq. (9) we now have

$$\begin{aligned} e_{\text{maser}} &= 1 - \frac{T_c}{T_h} \\ &= e_{\text{Carnot}}. \end{aligned} \tag{11}$$

It is interesting that we obtain e_{Carnot} for our maser heat engine without any of the usual hassle of expansion and compression strokes, etc.

Endnotes 7

"Frustrated by the deep mysteries of the second law, James Clerk Maxwell dreamed up an imaginary being designed to help him test the laws of thermodynamics. This being began life as his own private creature, but soon escaped and began to haunt the house of science ... It has teased physicists for more than a century and a quarter. Physicists do not know how to cope with it, so they usually just try to ignore it – but it refuses to go away. Its name is Maxwell's Demon."

Hans Christian Von Baeyer

E7.1 The Stern-Gerlach/Maser, Quantum Heat Engine

Science is wonderful. Strong arguments and emotional disagreements between scientists are common but usually short-lived, unlike many other fields of endeavor, such as politics and philosophy. With the publication of a new result comes the inevitable questioning from peers. Soon new calculations or experiments resolve the matter. In the process, everyone wins. For it is not an egotistical question of who was right and who was wrong. What matters is the discovery of new science.

The content of chapter 7 and these associated endnotes are cases in point. My interest in this problem was the result of an argument on the streets of Tel Aviv with two of my favorite colleges (Yakir Aharonov and Herbert Walther). One result of that argument is the enclosed excerpt from my Physical Review Letters paper titled "Extracting Work from a Single Thermal Bath via Quantum Negentropy." The main theme of the paper is that we can run a single quantum heat engine (e.g. a maser) by using information (about the atoms position) instead of a lower temperature entropy sink. The enclosed reprint and the following figure spell out the idea.

In the above figure, the atomic spin states are the working fluid, a thermal radiation source provides the energy and the atomic spins drive a maser field producing useful work. Thus, we are changing thermal light into coherent light, i.e. we are producing useful work. However, instead of using a low-temperature entropy sink, the center-of-mass degrees of freedom provide a source of negentropy (or information) which allows the QHE to operate for a finite number of cycles.

The negentropy heat engine does not violate the second law. What it does is to drive a maser engine by using quantum information, i.e., quantum negentropy that serves as an effective entropy sink. As in Fig. 7E-1, to complete a cycle, one needs to recombine the two beams. This recombination results in a doubling of the probability packet's width on each cycle. This yields in a state of increased disorder, or equivalently decreased negentropy, after each cycle. After N cycles, the width of the probability packet increases from $L = \ell_0$ to $L = 2^N \ell_0$ and at some point we have effectively lost the position information.

The Demon and the Quantum, Second Edition. Robert J. Scully and Marlan O. Scully

ISBN 978-3-527-40983-9

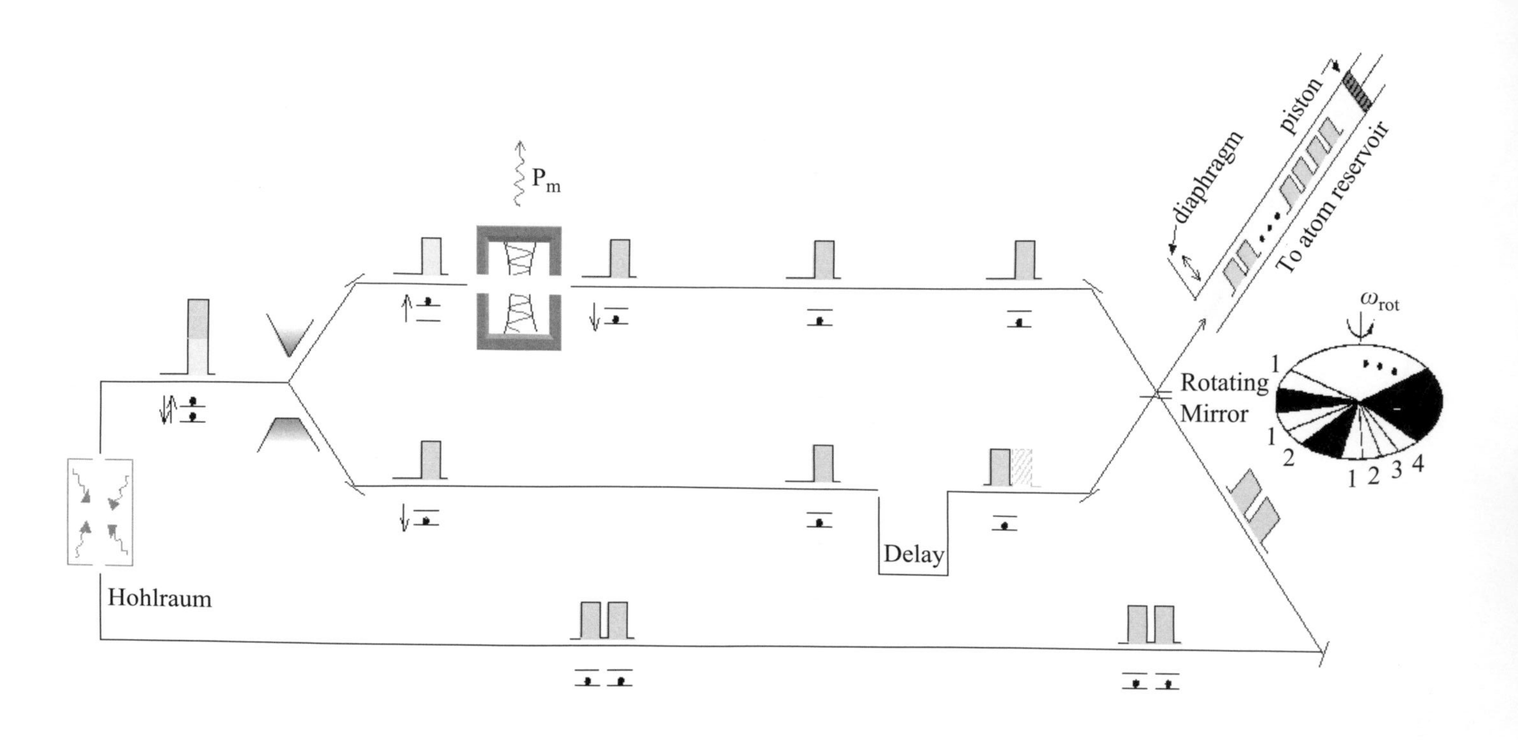

P_m
diaphragm
piston
To atom reservoir
ω_{rot}
Rotating
Mirror
1
1
2
1 2 3 4
Delay
Hohlraum

We then close the diaphragm, trap the atom, and isothermally compress it. The energy needed to regenerate the initial sharp packet of length ℓ_0 is simply the $\int P\,dV$ work of Fig. 7E-1. This work is $\Delta W = kT \log\left(\frac{2^N \ell_0}{\ell_0}\right) = NkT \log 2$.

Please note, thermodynamics obtains this result without information arguments. As before, there is an interesting parallel between this negentropy cost and Bennett's resolution of Maxwell's demon paradox on the basis of information theory [2]. Specifically, erasing an N bit memory register means converting a macrostate corresponding to 2^N possible microstates of the register to one of a single microstate. The "cost" for resetting each bit of information is $kT \log 2$ of isothermal compression work; hence, the energy needed to erase an N bit register is $kT \log 2^N$.

In lecturing about a problem involving a similar engine, which uses information to produce work, Feynman came to the same conclusion. Actually Feynman's lectures preceded my paper by over a decade but I did not come across Feynman's nice information theory lectures until preparing these endnotes! Entropy rejection via isothermal compression is closely related to, in fact equivalent to, information restoration. At this point in his lecture on information theory, Feynman said "Isn't it wonderful that [this isothermal compression work] has anything to do with what we are talking about!?"

E7.2 Making an Engine Operate on Information

The following Physical Review Letter is a precursor to the Stern-Gerlach quantum heat engine of Chapter 7. The main point in this paper is the utilization of information to run an engine for many cycles. That is, we do not need to restore the atomic position after each cycle.

Fig. 7E-1 In the negentropy heat engine, a thermally populated two-level atom, with a probability packet of width ℓ_0, undergoes state selection by a Stern-Gerlach apparatus. The excited atom passes through the resonant maser cavity and emits a photon. An atom in the ground state undergoes a time delay (note this delay is increased on each cycle), needed in order to recombine the beams by the rotating mirror. The mirror has alternating transparent and reflecting sectors and it is timed so that the upper beam gets transmitted and the lower one gets reflected. After the two beams are recombined by the mirror, the probability packet has twice its initial width ($2\ell_0$). Finally, the atom is directed back to the hohlraum and recycled. This can be repeated for $N \sim 20$ times before the atomic probability packet information is "scrambled" such that the atom is not being properly timed with the rotating mirror. Then it passes into the cylinder-piston "atomic reservoir" and, at this time, the diaphragm will be closed and the isothermal compression carried out.

Extracting Work from a Single Thermal Bath via Quantum Negentropy

Marlan O. Scully
Department of Physics and Institute for Quantum Studies, Texas A&M University, College Station, Texas 77843 and Max-Planck-Institut für Quantenoptik, D-85748 Garching, Germany
(Received 5 March 2001; published 12 November 2001)

Classical heat engines produce work by operating between a high temperature energy source and a low temperature entropy sink. The present quantum heat engine has no cooler reservoir acting as a sink of entropy but has instead an internal reservoir of negentropy which allows extraction of work from one thermal bath. The process is attended by constantly increasing entropy and does not violate the second law of thermodynamics.

DOI: 10.1103/PhysRevLett.87.220601 PACS numbers: 05.70.—a

The principle of detailed balance, so admirably applied by Einstein to the discovery of stimulated emission and derivation of the Planck distribution, is a mainstay of statistical physics. However, recent studies have shown how to break emission-absorption symmetry, yielding lasers operating without inversion and pointing the way to coherent control of thermodynamic processes. In fact, the original state selective masers (SSM) operated by sorting hot (spin up) from cold (spin down) atoms in order to get maser action from a thermal spin distribution.

The present work is an outgrowth of previous work, where it is shown that the internal "spin" states of an atom can be cooled to absolute zero via a SSM scheme, as in Fig. 1(b). We here show how this can be extended to design a quantum heat engine (QHE) based on cycling a single atom through a micromaser cavity many times.

In classical heat engines useful work is produced by drawing energy from a high temperature source and depositing entropy in a low temperature entropy sink. Specifically a working fluid, such as steam, draws energy from a boiler, does work on a piston, and deposits entropy in the cooling water. In the present QHE the atomic spin states play the role of the working fluid, a blackbody holhraum is the energy source, and the atomic spins drive a maser field producing useful work. However, there is no lower temperature entropy sink in the present QHE. Instead the atomic center of mass (c.m.) degrees of freedom are used to provide a source of negentropy which allows the QHE to operate for a finite number of cycles.

:
:
:

The one-atom QHE depicted in Fig. 1 is based on the fact that by using a carefully prepared two-level atom we can cyclically extract maser energy from a single thermal reservoir. In Fig. 1 we see a pulsed atom beam passing through a heat bath consisting of a hot microwave cavity in thermal equilibrium, i.e., a blackbody hohlraum at temperature T. The c.m. motion is essentially unaffected by the radiation heat bath, but the $|b\rangle$ spin state is changed to a thermal mixture.

It is important to emphasize that the atom undergoes state change only when in the various cavities, since the atom-field coupling can be made much stronger inside the cavity than outside. This is the central theme of cavity QED, wherein we routinely assume the atomic states are very long lived outside the cavity. As discussed in the figure caption, the atom passes from the hohlraum into an SGA where it is deflected into the two paths determined by the magnetic quantum numbers $m_z = +1, 0$. Please note that the SGA involves a conservative potential and does no work; i.e., the atom leaves the SGA with the same c.m. energy it had on entering. As depicted in Fig. 1(a) an atom in the $|a\rangle$ state passes into the high-Q maser cavity and its spin energy is transferred to the field by stimulated emission.

:

:

:

The atom is now recycled through the hohlraum. But now the c.m. wave packet has twice the width. Hence, for an atom passing on the L path we must increase the path length by $2l_0$ so that the U and L atomic pulses are again totally separated; see Fig. 1(d). We continue in this way, doubling the packet width on each cycle, until the atomic wave function fills the apparatus. Then it is no longer possible to recombine the beams by the time dependent mirror.

:

:

:

The similarities and differences with our toy QHE and the Maxwell demon problem are interesting. In particular, our QHE has much in common with the Szilard single atom engine. For example, it is interesting that the estimate $W_{\text{prep}} = NkT \ln 2$ given above is in agreement with the Szilard-Bennett result obtained on the basis of the theory of computing. However, no measurement is made in the operation of our QHE. The present analysis does, however, focus on the quantum information, i.e., quantum negentropy, associated with the atomic c.m. position.

The interplay between the second law of thermodynamics and quantum mechanics has a long history. The pioneering work by Ramsey proved that the Kelvin-Planck statement of the second law had to be revised when (quantum) negative temperatures were introduced. The fact that the laser, driven by three-level atoms, could be viewed as a kind of quantum heat engine was pointed out some time ago. However, the present two-level-state selection engine has more in common with the 1929 classical one-atom engine of Szilard.

E7.3 Entangling Maxwell's Demon with Seth Lloyd

In the main text, Rob teaches us how to resolve the Maxwell demon paradox by using a Stern-Gerlach device to sort hot from cold atoms. Since the SG deflector is a quantum device we can correctly say that he used quantum mechanics to resolve the demon challenge to the second law.

However, there is another very pretty way to use quantum mechanics to resolve the demon paradox which we shall sketch here. This is the "entanglement" approach to the problem due to Seth Lloyd. As it was originally presented (in the *Physical Review A*) the problem was carefully laid out using two particle correlation or entangled states. Here we shall use Lloyd's main idea, i.e., two particle entanglement, but we will apply it to running a maser quantum heat engine. The abstract to Loyd's paper reads:

PHYSICAL REVIEW A VOLUME 56, NUMBER 5 NOVEMBER 1997

Quantum-mechanical Maxwell's demon

Seth Lloyd*
d'Arbeloff Laboratory for Information Systems and Technology, Department of Mechanical Engineering, Massachusetts Institute of Technology, MIT 3-160, Cambridge, Massachusetts 02139
(Received 8 November 1996; revised manuscript received 26 March 1997)

A Maxwell's demon is a device that gets information and trades it in for thermodynamic advantage, in apparent (but not actual) contradiction to the second law of thermodynamics. Quantum-mechanical versions of Maxwell's demon exhibit features that classical versions do not: in particular, a device that gets information about a quantum system disturbs it in the process. This paper proposes experimentally realizable models of quantum Maxwell's demons, explicates their thermodynamics, and shows how the information produced by quantum measurement and by decoherence acts as a source of thermodynamic inefficiency.

Let us consider how we can use Seth Lloyd's idea to run our maser. Suppose we have a heat oven which the atoms (type 1 atoms of Fig. 7E-2) pass through emerging

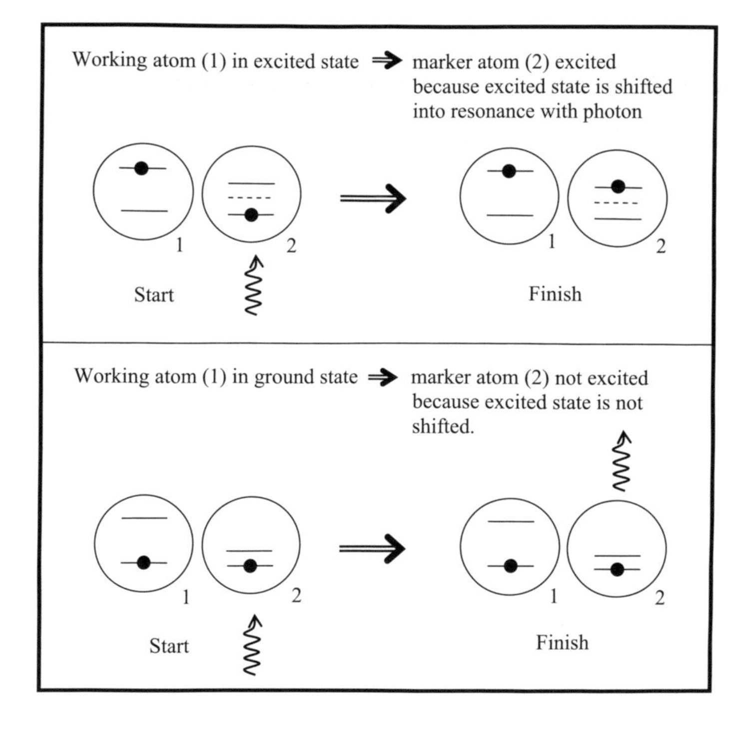

Fig. 7E-2 Showing how the state of the maser atom (type 2) is excited by incident microwaves (or not) depending on the state of the nearby type 1 atom.

in a hot/cold mixture as in Chapter 7. There we proceeded to separate hot (excited) atoms from cold (ground state) ones by the Stern-Gerlach apparatus. Now, following Lloyd, we introduce marker atoms (type 2 atoms) which we place near the type 1 atoms. Now the type 2 atoms start off in the ground state. We also have a beam of microwaves which are tuned to excite the type 2 atom if the type 1 atom is in the upper level but leave it alone if the type 1 atom is in its ground state. This is possible since the excited state type 1 atom will change the binding energy of the type 2 atom but if atom 1 is in the ground state it will not affect the marker, type 2, atom. This is depicted as follows.

So now we have a kind of Szilard situation. We only need look at the type 2 marker atom to tell if the working type 1 atom is excited or not. If the marker atom is in the upper level, then the type 1 atom is excited and we put the working atom in the maser cavity and gain useful work by stimulated emission. If the marker atom is not excited, we don't put the working, type 1, atom in the cavity.

We assume the excitation energy of the type 1 atom, ε_1, is greater than that of the type 2 atom ε_2. Hence we get a gain of useful work $\varepsilon_1 - \varepsilon_2$.

However, we must now reset the type 2 atom back to the ground state. That is, we must extract entropy $S = k \log 2$ from the marker atom, just as was the case in the single atom Szilard engine. Since there is only one temperature involved (the temperature of the oven) the erasure work required is $TS = kT \log 2$.

To summarize:

1. By using a marker atom we can tell if the working atom is hot or cold.
2. Acting on this information, we use hot atoms to run our maser heat engine, but ignore cold ones.
3. However, just as in Szilard's engine, we must erase the information stored in the marker atom in order to complete the thermodynamics cycle.

The bottom line is that we see how it is that a careful analysis of the demon problem requires us to spend more energy operating our measuring device (i.e., erasing its memory) than the useful work we get out. There is, therefore, no violation of the second law.

E7.4 The Cover Story

The beautiful cover of Rob's book depicts two optical fibers feeding into one via a "Y" junction. The two input fibers are a dull red and the output fiber is a bright red indicating that there are more photons in the right hand (output) fiber than the input. In fact, if there are N photons in each of the left hand (input) fibers we might expect there would be $2N$ photons in the right hand (output) fiber. Here is why that is wrong.

If the temperature of the light is very high in the sense that the thermal energy kT is much larger the photon energy ε, then the number of photons is given by

$$N = \frac{kT}{\varepsilon}.$$

This is an intuitively reasonable result. It says that if the thermal energy in the monochromatic fiber photon field is kT, and if the energy per photon is ε, then the number of photon packets is just the total thermal energy divided by the energy per photon.

Therefore, if we could double the number of photons, by feeding N photons from each input branch, we would have

$$T_{\text{in}} = \frac{N_{\text{in}}\varepsilon}{k},$$

and

$$T_{\text{out}} = \frac{2N_{\text{in}}\varepsilon}{k}.$$

That is, we would have doubled the temperature of the right hand (output) field compared to the input field we started with.

However, if we could do this we could run an engine between the hot $(2T_{\text{in}})$ and cold (T_{in}) "baths" of incoming and outgoing photons. This would violate the second law of thermodynamics.

Hence, the logic goes, we must not be able to combine the two photon beams in the way indicated on the cover. As you might expect, we could make such a beam combiner if we had a Maxwell demon. More on this problem is found in the nice article by Jackel and Tomlinson reprinted in the next section.

E7.5 Why It Is Not Possible to Combine Two Incoherent Photon Beams (Without Maxwell's Demon)

The fact that the merging of two incoherent photon beams into one is impossible is well known to those who know it well. Jackel and Tomlinson write the following charming account of the problem for those who don't "know it well."

The lossless single-mode waveguide combiner

BY J. L. JACKEL AND W. J. TOMLINSON

When a symmetric single-mode waveguide y-branch 50:50 power splitter is used as an optical combiner, there is an inevitable loss of at least 3 dB. Why this happens may not be intuitively obvious. This short tutorial shows why this loss cannot be avoided. We present analyses based on the second law of thermodynamics, on the reversibility of optics, and on the evolution of the optical waveguide modes propagating in this structure. We also show how modifications of the y-branch structure allow its use as a lossless optical combiner, but only under very specific conditions that do not conflict with any of the above explanations.

The optical waveguide y-branch is a basic building block for optical waveguide components. The simplest and one of the most valuable of the components based on the y-branch is the 50:50 power splitter. A symmetric y-branch acts as a 50:50 optical power splitter, and practical devices are possible with measured total losses of the order of 0.1 dB. It is tempting to consider using such a y-branch power splitter as a combiner. If it can split an optical signal into two equal components, one might expect that it can be used in reverse to combine incoming signals into a single waveguide. The (illusory) "lossless" single-mode waveguide combiner is such an attractive device that it is "invented" frequently. Its appeal is sufficiently seductive that many enthusiasts need more than one explanation of why the lossless combiner is an "integrated optical illusion."[1]

In reality, a symmetric y-branch can combine signals, but only at the cost of a loss of at least 3 dB. We show why this loss is inevitable, using both extremely general arguments based on thermodynamics, and on the reversibility of optical components, and an analysis based on the evolution of optical waveguide modes. We also show that the same arguments preclude use of any wavelength- and polarization-independent symmetric power splitter as an optical power combiner without the same 3-dB minimum loss. Real devices always have imperfections that result in additional or excess losses, but in this note we only consider intrinsic losses.

A waveguide y-branch, shown in Figure 1a, consists of a single-mode waveguide (which might be a fiber) which is split symmetrically (or asymmetrically) to become two single-mode guides. Light entering the single guide passes through the branch and is split into two equal parts (in the symmetric case), each of which has half the input intensity. For our purposes the shape of the splitting region is unimportant, although in practice it determines the excess loss of the branch.

Thermodynamics

The second law of thermodynamics provides us with a simple way to see that a 3-dB loss is inevitable when we use a symmetric y-branch as a combiner. One general statement of the second law is that a net transfer of energy from a low temperature region to a high temperature region is not possible, unless work is performed on the system. Indeed, there can be no net transfer of energy from one region to another having the same temperature, unless work is performed.

To see how this applies to the y-branch, consider Figure 1b. Suppose we place a light bulb at the entrance of the single guide, A, and an identical light bulb at the entrance of one of the branches, B. We assume that an equal amount of light is coupled into each guide. Now, consider the amount of light from A that reaches Point B. It must be one-half the light that is coupled into Guide A, since this is a 50:50 power splitter. If all the light from Branch B could be passed into Branch A, then twice as much light would be sent from B to A as is sent from A to B. This would violate the Second Law. The only way that the behavior of the waveguide branch can be consistent with thermodynamics is that exactly one-half of the light in Branch B be coupled into Branch A.

Carrying this argument one step further, we see that if the branch is not symmetric the fraction of light coupled from

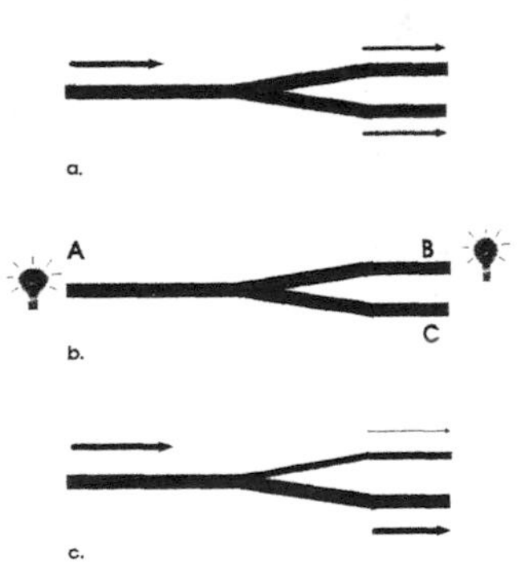

Fig. 1 Waveguide y-branches: (a) A symmetric branch, used as a power splitter; (b) A symmetric branch with light sources used for a thermodynamics analysis (see text); (c) An asymmetric branch.

Branch B to Branch A must still be identical to that coupled from Branch A to Branch B. Thus, in an asymmetric y-branch (Fig. 1c) which couples, for example, only 5% of the light into one of the branches (as might be needed in a waveguide tap), only the same 5% of light could be coupled from that branch to the single guide. However, in this case, 95% of the light from the other branch would be coupled into the single guide. (For simplicity, in the balance of this paper we only consider symmetric y-branches, but the same principles and explanations can be applied to asymmetric branches.)

While this argument is correct and gives us additional information about the behavior of asymmetric branches, it may not be completely satisfying. Thermodynamics, after all, was devised to describe the behavior of heat engines, not light. In addition, this argument fails to address one pressing question: If half the light is lost, where does it go?

Reversibility of optics

A basic principle of optics is that passive optical systems are time reversible. If a certain input to a component produces a certain output, then sending that output back through the component must reproduce the original input. How does this apply to the y-branch splitter? We know that if a certain intensity of light, I, enters the single guide, half that intensity, i.e., $I/2$, will exit each of the two output branches. To reverse the behavior of the branch, we cannot introduce light into only one of the output branches; both of the output branches must have equal amounts of input light, $I/2$, to reproduce the output state.

One might then expect that sending light with intensity $I/2$ into both of the output guides would reproduce the original input. However, this is not necessarily the case, because we must also consider the phase and the polarization of the light. The outputs of the branch have well-defined relative phases and polarizations, as well as equal intensities. Only by introducing light with phases, polarizations, and intensities matching the output of the splitter can we make the combiner act as the inverse of the splitter. If we do this, the output not only reproduces the original input intensity, I, but the original optical field amplitude, which is proportional to $I^{1/2}$. From symmetry, each of the two input signals must then be contributing half of that field amplitude, or $I^{1/2}/2$. If we shift the relative phase of the input fields by (π, these two contributions will then be out of phase and cancel, resulting in zero output. For uncorrelated input signals the relative phase will be a rapidly varying function of time, and the time-averaged output intensity will be the average of these two extremes, or $I/2$ which corresponds to a 3 dB loss. If one of the input signals is eliminated, the resulting output intensity will be the square of the field contributed by the remaining input, or

$$[\sqrt{I}/2]^2 = I/4$$

which also corresponds to a 3 dB loss. (In a similar manner, one can also analyze the effects of variations of the polarizations of the input signals.) However, the principle of optical reversibility still does not explain what happens to the light that is lost. To understand this, we need to look explicitly at the behavior of the optical modes in a branch.

Evolution of modes in a branch

Figure 2a shows the optical mode propagating in the waveguide branch. In the single-guide region of the branch only one mode can propagate. As light enters the waveguide branch, that mode becomes wider. Even as it

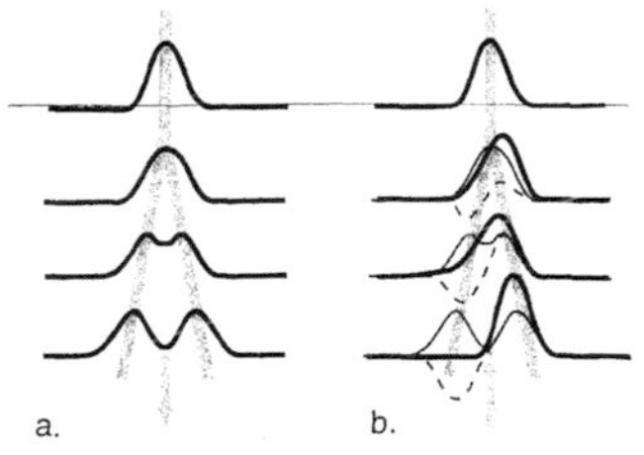

Fig. 2 Evolution of the waveguide modes in a symmetric y-branch. (a) Used as a power splitter. (b) Used as a combiner, with input to only one branch. The black curves show the total optical field *amplitude*. The grey curves show the amplitude of the symmetric mode, and the dashed curves are the asymmetric mode.

reaches a region which can support a second (asymmetric) mode only the original mode propagates, since nothing has occurred to cause the second mode to be excited. In the region where there are two single mode guides, the fundamental mode has changed shape; half of the light intensity $[1/(2)^{1/2}$ of the amplitude$]$ is in each of the guides, and the phase is equal in the two branches. This is generally referred to as the symmetric mode.

Figure 2b shows the same branch used as a combiner, with light entering only one of the branches. Now the distribution of light cannot be described solely in terms of the symmetric mode. Rather, it is the sum of the symmetric and the antisymmetric modes. What happens when the light reaches the branching region? The symmetric mode passes through it, but the antisymmetric mode cannot propagate in the single-mode region. As it moves through the branch, it becomes more and more spread out, as the waveguide confines the light less and less. At a certain point, the waveguide can no longer confine this mode. At this point we say that the mode is cut off, and the light is radiated into the substrate. This way of looking at the behavior of the branch explains both why there is a 3 dB loss, and where the lost light goes.

(Note that this argument depends on the combiners having a single-mode output guide. If the two branches were combined into a guide having two modes, then all the light could pass through it. However, if the two-mode section was subsequently converted to a single-mode guide, the asymmetric mode would be lost. There are many guided wave components which use two-mode sections to perform certain functions, but discussing them is beyond the scope of this note.)

Some generalizations, and one specific example: Nothing in the discussion so far has depended on the details of the waveguide construction. In fact, the arguments based on thermodynamics and on the reversibility of optics did not even assume that the power splitter consisted of a waveguide branch. The same behavior will occur if the power splitter is based on a waveguide directional coupler, beam-splitters, or mirrors. Any component that takes an arbitrary signal and separates it into two equal signals will inevitably have a 3 dB loss when used as a combiner.

However, there are useful combiners which appear to evade this requirement. One of the most obvious is the wavelength splitter, which takes an input containing light at two different wavelengths and routes these two components to different output ports. If this splitter is used in reverse, it can combine the wavelengths it separates, without loss. Does this invalidate the previous arguments?

In fact, there is no contradiction. Suppose only one wavelength enters the splitter and goes to the appropriate output. If we time reverse the splitter, this wavelength enters where it would have exited. In this case, the light entering what had been one of the outputs of the splitter will all go to the input, but only if it is of the correct wavelength. The splitter can thus operate as a lossless combiner, but only if the light entering each port has the correct wavelength for that input port and all the wavelengths are distinct. Light entering an inappropriate port is lost.

For the same reasons, a structure which splits incoming light into two orthogonal polarizations can combine those polarizations without loss, but only if the light enters the port appropriate for its polarization. Such splitters, like wavelength splitters, can be constructed in waveguide form or using bulk optics. However they are made, the same principles of reversibility apply.

Conclusions

We have seen that it is not possible, even in principle, to construct a symmetric single-mode optical power combiner which does not have an intrinsic 3 dB loss. However, it is possible to construct lossless wavelength or polarization combiners. Understanding this fundamental, though non-intuitive, behavior may be as important for the design of systems as it is for components.

Notes

1. To the best of our knowledge, the use of the term an "integrated optical illusion" to refer to the lossless single-mode waveguide combiner was coined more than 10 years ago by Herwig Kogelnik. Despite active encouragement by one of us (WJT), he never found the time to write the paper. In the present paper we have appropriated Kogelnik's term, and we have aspired to present the fundamental concepts as clearly as he would have done.

J. L. Jackel and W. J. Tomlinson are at Bellcore, Red Bank, N.J.

Endnotes 8

$$\Delta x \Delta p \geq \hbar$$

"Einstein said that if quantum mechanics is correct the world is crazy. Einstein was right, the world is crazy."

Daniel Greenberger

E8.1 The Light Side of Waves: Sines and Cosines Again

Wave behavior is all around us. Our audio capabilities bears mute (anti-mute?) testimony to this fact. Music, from Pythagoras to Mozart touches the soul. What is this music thing that brings us so much pleasure? What exactly is sound and how does it propagate?

Let us begin to answer these questions by thinking of what happens when our ear perceives a sound impulse, e.g. a rifle shot. The molecules in the air are, to a good approximation, held in place (by collision with neighboring molecules), at least for "short times" of order a few milliseconds.

Now, suppose we have a drumhead that we strike at time $t = 0$. Then the air molecules will abruptly move to the right, striking the adjacent molecules; there molecules will transfer the impulse they have received to those further "down the line" and so on as in Fig. 8E-1.

Another example of sound propagation is the motion of atoms in a solid rod following a hammer blow at one end of the rod. The atoms are held in place by the electronic forces between them, which we represent as springs in Fig. 8E-2. The solid rod is analogous to the air/gas example of Fig. 8E-1.

We can understand and explain sound waves better by considering wave propagation as an extension for the pulse propagation of Figs. 8E-1 and 8E-2. Consider another model of atoms held in place and weakly coupled to their nearest neighbors as in Fig. 8E-3. Here we represent our rod by replacing each atom by a pendulum. The dotted lines represent weak springs coupling the "atoms."

Next we consider one pendulum alone. At time $t = 0$, we displace it by a small amount; thereafter, it will oscillate back and forth about the equilibrium position as in Fig. 8E-4.

The Demon and the Quantum, Second Edition. Robert J. Scully and Marlan O. Scully

ISBN 978-3-527-40983-9

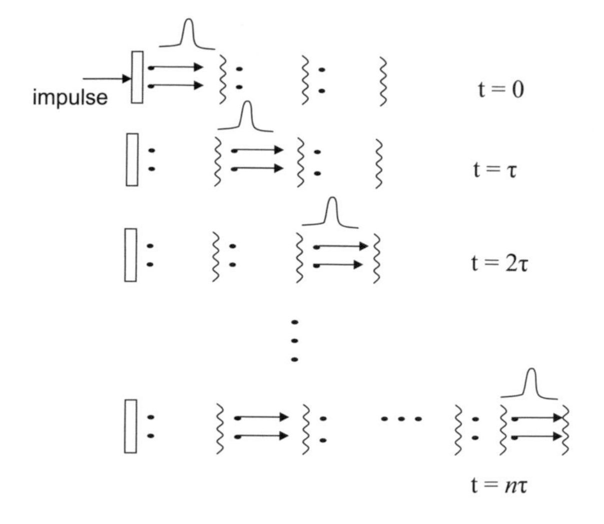

Fig. 8E-1 The motion of air molecules corresponding to a pulse of sound from a drum which is struck at time $t = 0$.

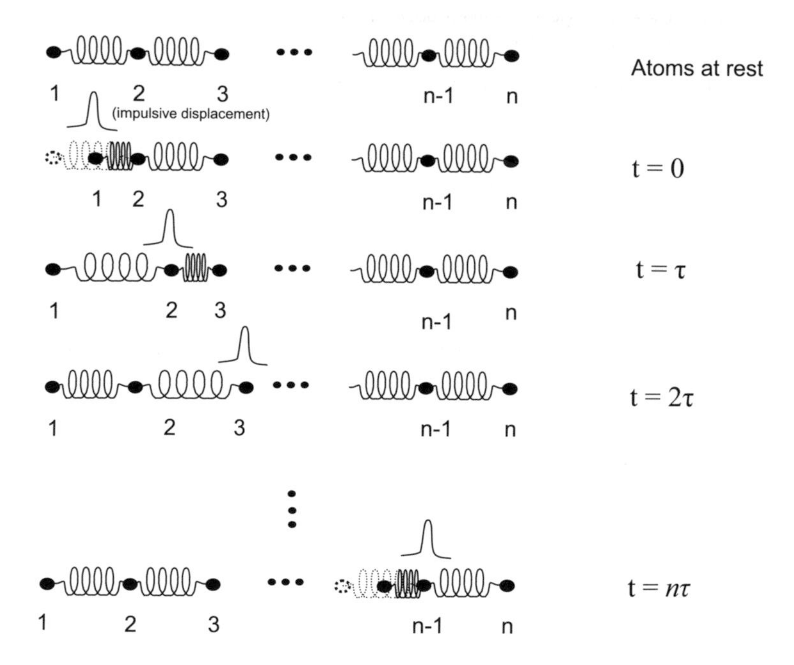

Fig. 8E-2 Atoms displaced by an impulse at the left transfer the momentum imparted to them to atoms on the right. In this way, sound moves through the rod.

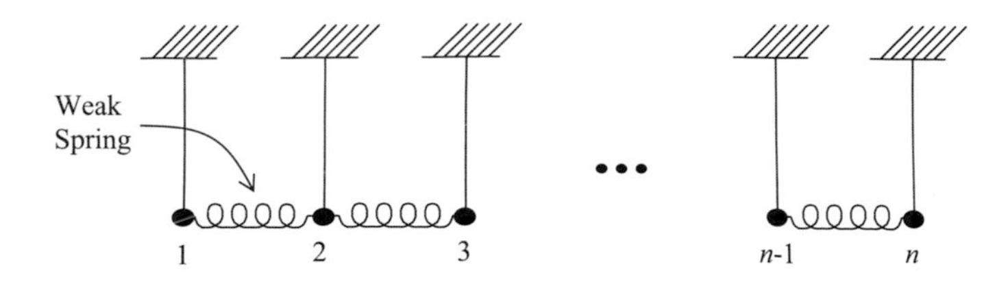

Fig. 8E-3 Model of a solid rod in which the atoms are thought of as pendulums which in their unperturbed state represent solid at temperature $T = 0$. The "atoms" are coupled by weak springs so that a pulse of sound travels down the rod as in Fig. 8E-2.

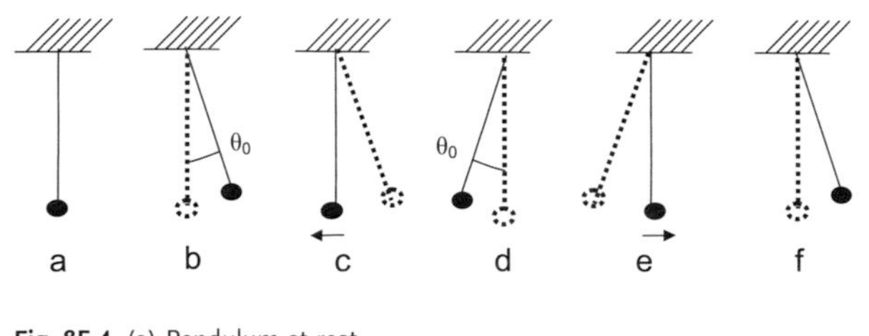

Fig. 8E-4 (a) Pendulum at rest.
(b) displaced by angle $\theta = \theta_0$ (no velocity) $t = 0$
(c) swings to down position $\theta = 0$ $t = \tau$
(velocity to left).
(d) swings to maximum position $\theta = -\theta_0$ $t = 2\tau$
(no velocity)
(e) swings to down portion $\theta = 0$ $t = 3\tau$
(velocity to right)
(f) returns to $\theta = \theta_0$ (no velocity) $t = 4\tau$

Please note that the displacement in the horizontal x direction is given by

$$x = x_0 \cos \theta, \tag{1}$$

where x_0 is the maximum displacement. Now, the angle θ can be related to the time by writing $\theta = 90^\circ \frac{t}{\tau}$, so when $t = 0$, $\theta = 0^\circ$, and when $t = \tau$, $\theta = 90^\circ$ etc. However, we prefer to measure angles in terms of radians, defined as the fraction of an arc A, of a circle of unit radius, subtended as in Fig. 8E-5.

Written in radians, Eq. (1) reads

$$x = x_0 \cos\left(\frac{\pi}{2}\frac{t}{\tau}\right). \tag{2}$$

It is natural to think in terms of complete cycles T such that $\tau = T/4$ so that we have the displacement of our pendulum as

$$x = x_0 \cos\left(2\pi \frac{t}{T}\right), \tag{3}$$

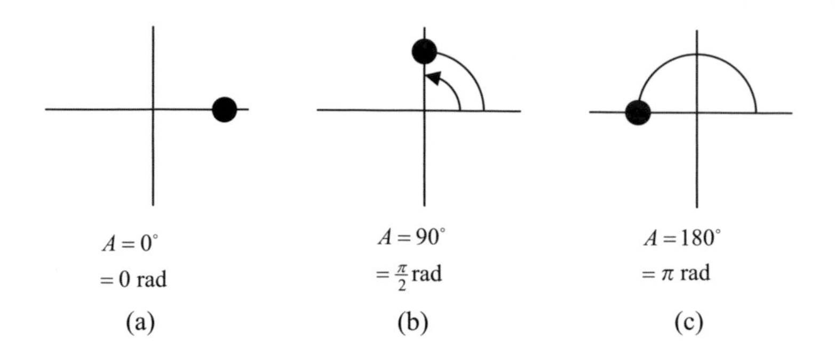

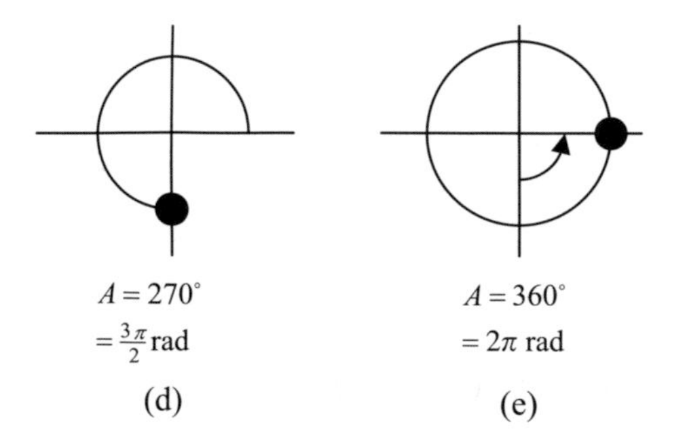

Fig. 8E-5 Figures illustrating angular measure in degrees and radians.

or

$$x = x_0 \cos(\omega t), \tag{4}$$

where we define the frequency as $\omega = 2\pi/T$ radians per second.

Next, let us consider how a single pendulum set in oscillation will communicate with atoms (pendula) down the rod. In fact, an impulse at $t = 0$ will be coupled to another (call it the ith pendulum) at some later time, t_i, so that its state of oscillation will be given by

$$x_i = x_0 \cos \omega(t - t_i). \tag{5}$$

Now the time at which the ith atom is effectively displaced is given by its position, call it z_i, and the velocity of the impulse, v by $vt_i = z_i$. So that we may write Eq. (5) as

$$x_i = x_0 \cos\left(\omega t - \frac{\omega z_i}{v}\right). \tag{6}$$

We pass to the continuum limit by noting that $x_i = x_i(t, z_i)$ and since the ith atom is uniquely determined by its position so we drop the subscript i and write $x_i(t, z)$. Finally, we write

$$x(t, z) = x_0 \cos(\omega t - kz), \tag{7}$$

where we have defined the wave number $k = \dfrac{\omega}{v}$. The wave number k is $2\pi/\lambda$ where λ is the wavelength, and k has dimension of $1/L$.

E8.2 The Wave Side of Light

It is a historical fact that Newton thought of light as a beam of particles. He reasoned that if light were wave-like, it would bend around corners. Such behavior was not observed (in his time), so he opted for a particle picture. Over a century later, Thomas Young did his famous two-slit experiment showing the interference behavior discussed at length in the text.

Here we carry out a little calculation to show how the interference pattern arises. We begin by writing the wave field from two sources A and B as $E_a = E_0 \cos(\omega t - kr_a)$, $E_b = E_0 \cos(\omega t - kr_b)$, where E_0 is the strength of the light wave and r_a is the distance from A to the detector while r_b is the distance from B.

Consider the cases depicted in Fig. 8E-6 in which the detector is placed in different positions.

In order to see all this in a simple way, consider the case in which a short pulse is emitted at $t = 0$ and arrives at the detector at time $t_a = r_a/c$ from A and B. Then the fields at the detector are given by

$$E = E_a + E_b = E_0 \cos\left(\omega \frac{r_a}{c} - kr_a\right) + E_0 \cos\left(\omega \frac{r_a}{c} - kr_b\right). \tag{8}$$

Recalling that $\dfrac{\omega}{c} = k$, we have $E = E_0 \cos(0) + E_0(k(r_a - r_b))$.

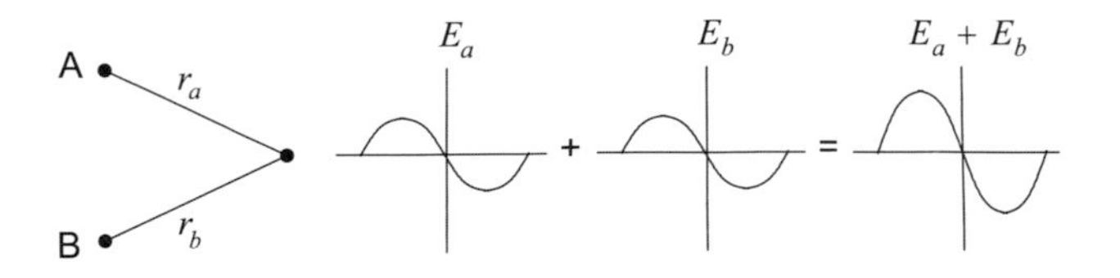

(a) Constructive interference when $r_a = r_b$ (Bright spot)

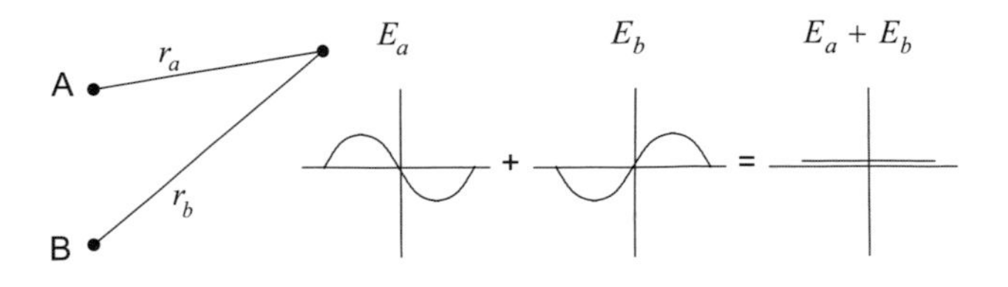

(b) Destructive interference when $r_b = r_a + \lambda/2$

Fig. 8E-6 Constructive (a) and destructive (b) interference occur depending on the relative distance from the source to the detector.

Hence, if $r_a = r_b$ (as in Fig. 8E-6a), we see that

$$E = E_0 + E_0 = 2E_0, \quad \text{Constructive interference} \tag{9a}$$

while if $r_a - r_b = \frac{\pi}{k} = \frac{\lambda}{2}$, we have

$$E = E_0 - E_0 = 0, \quad \text{Destructive interference} \tag{9b}$$

since $\cos[k(r_a - r_b)] = \cos \pi = -1$.

E8.3 Matter Waves

As discussed in Chapter 8, quantum particles, like electrons, neutrons, and atoms, etc., have both wave and particle attributes. However, the wave function for, say, an electron is different from the function describing sound or light waves. The functional form for electrons, light, and sound are given in Table 8E-1.

Table 8E-1 shows the salient difference between quantum (matter) waves and classical (sound) waves: classical waves are real while quantum waves are usually complex (i.e. must be described with the imaginary number i). The amplitude, $X(z, t)$, of a sound wave or the electric field, $E(z, t)$, of a radio wave have simple physical meaning. For example, the electric field can be measured by deflecting an electron or ion beam.

The complex quantum wave function $\psi(x, t)$ is a "probability amplitude" wave. As Max Born taught us, the absolute value of the wave function squared $|\psi(z, t)|^2 = \psi^*(z, t)\psi(z, t)$ is the probability that a particle detector placed at z will register a count at t. Dirac introduces the state vector $|\psi(t)\rangle$ which is the state vector for our quantum particle.

It is useful to compare the state vector for a particle with, say, its position vector as in Fig. 8E-7.

The eigenvectors of a particle correspond to the outcome of possible measurements, e.g., the projection onto the eigenvector $|r\rangle$ tells us the probability amplitude that the particle "is at" point $\vec{r}$ (or better will be found at point $\vec{r}$), that is $\langle\vec{r}|\psi\rangle = \psi(\vec{r})$. In particular the absolute value squared of $\psi(r)$,

Table 8E-1 Comparison of Classical waves (sound, light ...) and quantum waves (electrons, atoms ...) showing that the classical wave function is real but the quantum one is usually complex.

Physical Waves	Mathematical Wave Function
Sound	$X(z, t) = x_0 \cos(\omega t - kz)$
Light	$E(z, t) = E_0 \cos(\nu t - kz)$
Electrons	$\psi(x, t) = \psi_0 e^{-i(\nu t - kz)}$

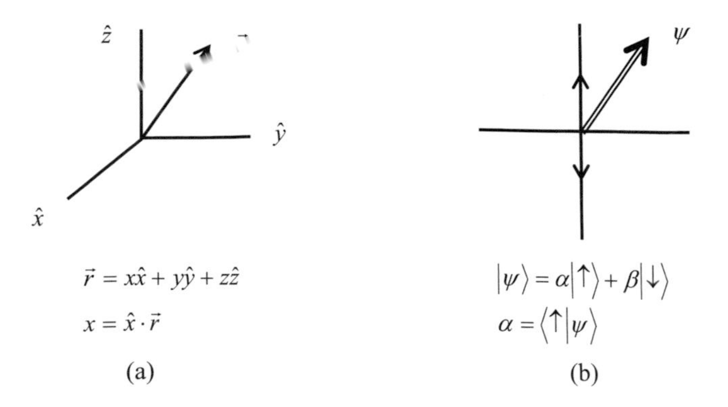

Fig. 8E-7 (a) The position vector of a particle is given in terms of the three unit vectors $\hat{x}$, $\hat{y}$, $\hat{z}$ and the projection onto those unit vectors. (b) The state vector for a particle having two eigen ("unit") vectors, up $|\uparrow\rangle$ and down $|\downarrow\rangle$ as in the case of the atomic magnets discussed at length in the next chapter. The projections onto the up and down directions are given by the inner products $\alpha = \langle\uparrow|\psi\rangle$ and $\beta = \langle\downarrow|\psi\rangle$ respectively.

$$|\langle r|\psi\rangle|^2 = \psi^*(r)\psi(r) = P(r) \tag{10}$$

is the probability density $P(r)$ that a particle will be found at $\vec{r}$ (if we put a detector there).

On the other hand, we may want to know how fast a quantum particle is moving, or what its momentum, $\vec{p}$ is. The eigenvectors $|\vec{p}\rangle$ provide us with the tools to answer this question. The "wave function" in momentum space is now given by

$$\langle\vec{p}|\psi\rangle = \psi(\vec{p}), \tag{11}$$

and the probability density of finding the particle to have momentum $\vec{p}$, $\mathscr{P}(\vec{p})$, is given by

$$|\langle\vec{p}|\psi\rangle|^2 = \psi^*(\vec{p})\psi(\vec{p}) = \mathscr{P}(\vec{p}). \tag{12}$$

E8.4 Two State Systems

An outline is given of the state vector treatment of three problems: (A) two slit interference, (B) beam splitter, and (C) atomic micromagnets.

A) The two source or two slit problem
The interference experiment of Fig. 8E-8 has the which-source representation of $\psi_1(r)$ and $\psi_2(r)$ and the fringe and anti-fringe representations (These are wave functions, not state vectors).

$$\psi_+ = [\psi_1(r) + \psi_2(r)]/\sqrt{2}, \tag{13a}$$

$$\psi_- = [\psi_1(r) - \psi_2(r)]/\sqrt{2}, \tag{13b}$$

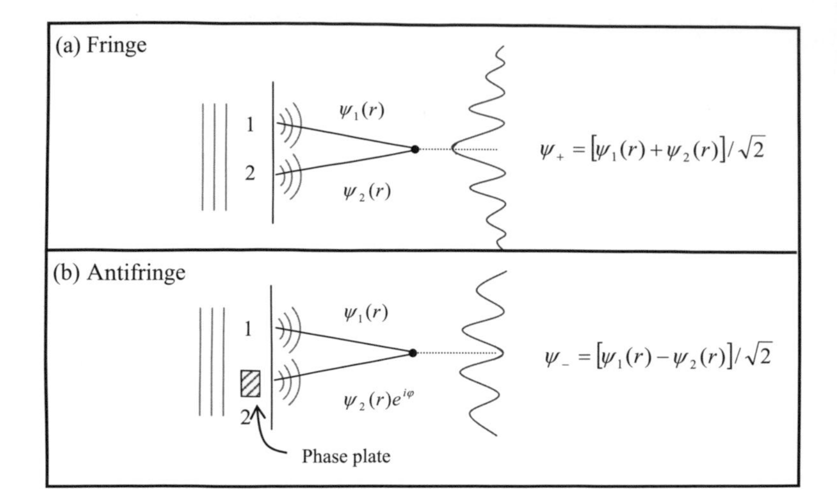

Fig. 8E-8 The which-path (which-source) states ψ_1 and ψ_2 are complementary to the fringe ψ_+ and ψ_- states. The ψ_- state is prepared by a phase delay plate, which puts a minus sign in front of the ψ_2 state to produce ψ_-.

B) Beam Splitter
Consider a particle (photon, atom, etc.) which is incident on a beam splitter from the top in momentum state $|p_\alpha\rangle$ and from the bottom with the momentum state $|p_\beta\rangle$. The beam splitter takes the state $|p_\alpha\rangle$ into the symmetric state

$$|s\rangle = (|p_\beta\rangle + |p_\alpha\rangle)/\sqrt{2}, \tag{14a}$$

$$|\bar{s}\rangle = (|p_\beta\rangle - |p_\alpha\rangle)/\sqrt{2}, \tag{14b}$$

as shown in Fig. 8E-9.

C) Two State Atomic Magnets
In Chapter 6, we focused on the two state (up, down) magnet. This state was historically associated with a spinning electron and it was found that the spin as either parallel to the magnetic field or anti-parallel as in Fig. 8E-10, thus the jargon "spin up" or "spin down", as

spin up: $|\uparrow\rangle$,

spin down: $|\downarrow\rangle$.

We could as well point the field along $\hat{x}$ direction and the spin is now going to be found to point along the $+x$ or the $-x$ direction. In terms of the $|\uparrow\rangle$ and the $|\downarrow\rangle$ states, we find that the $|+x\rangle$ and $|-x\rangle$ states are

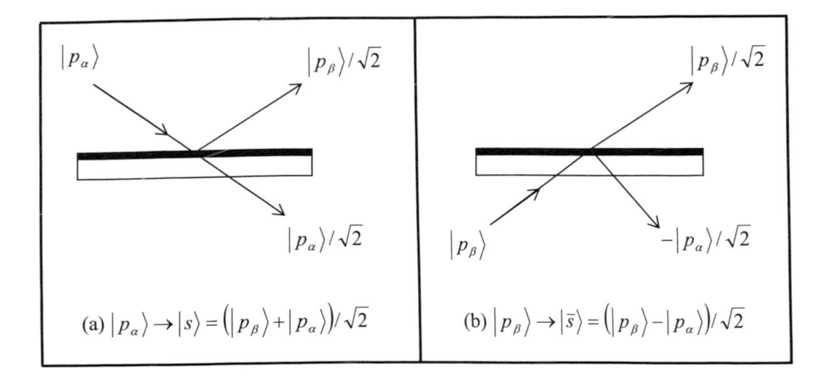

Fig. 8E-9 Orthogonal state $|p_\alpha\rangle$ and $|p_\beta\rangle$ are transformed such that $|p_\alpha\rangle$ goes to $|s\rangle$ and $|p_\beta\rangle$ goes to $|\bar{s}\rangle$ which must be orthogonal to $|s\rangle$.

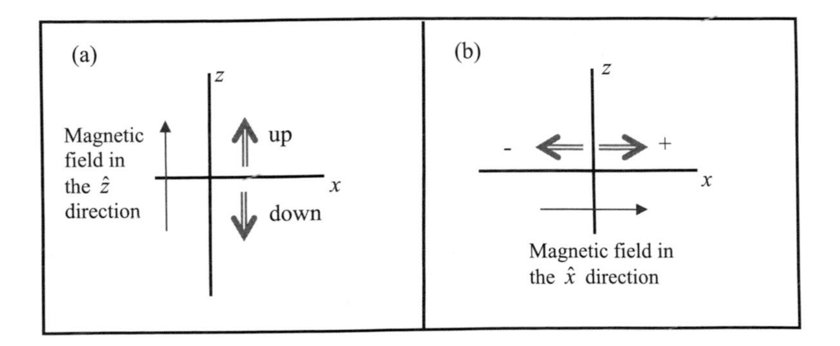

Fig. 8E-10 The two states of silver atom micromagnet correspond to choice of magnetic field alignment.

$$|+x\rangle = \frac{1}{\sqrt{2}}[|\uparrow\rangle + |\downarrow\rangle], \tag{15a}$$

$$|-x\rangle = \frac{1}{\sqrt{2}}[|\uparrow\rangle - |\downarrow\rangle]. \tag{15b}$$

E8.5 Single Slit Diffraction

Finally, we consider the two slits of Young's experiment moved closer and closer together until they merge to become one slit. The interference pattern that results is another example of wave behavior. The intensity distribution at the screen is now given by the simple relation

$$I = I_0 \frac{\sin^2 \alpha}{\alpha^2} \tag{16}$$

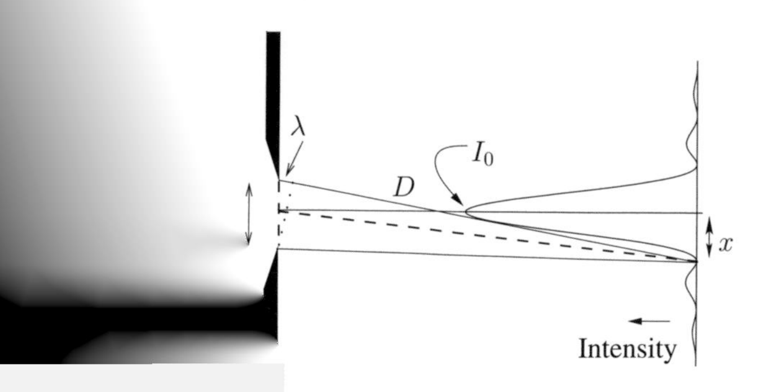

Fig. 8E-11 Single-slit diffraction. Light shining through a single slit produces the diffraction pattern shown above. Depending on the distance from the center of the screen, the light intensity undergoes a series of maxima and minima.

Where I is the intensity at some point x on the screen and I_0 is the intensity of the incident wave. The quantity α is given by

$$\alpha = \frac{\pi a x}{\lambda D},$$

where as in Fig. 8E-11,

a = slit width,
x = distance on screen from the center line,
λ = wavelength,
D = distance from slit to screen.

Endnotes 9

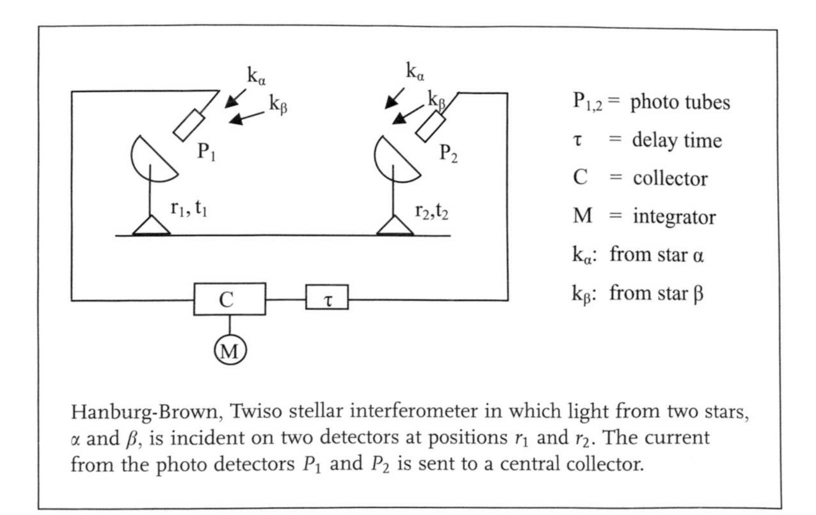

Hanburg-Brown, Twiso stellar interferometer in which light from two stars, α and β, is incident on two detectors at positions r_1 and r_2. The current from the photo detectors P_1 and P_2 is sent to a central collector.

> "The essence of the trick used by Hanbury Brown, and Twiss was to detect the signals first. This experiment is quite different in nature from the Young interference we described in Chapter 7 because the present experiment uses two photodetectors (not one as in Young's experiment)."
>
> Adapted from the "Les Houches" lectures of Nobel Laureate Roy Glauber

E9.1 Quantum Eraser Review Article

The following excellent review of the quantum eraser idea and extensions (e.g., to particle physics) by Aharonov and Zubairy is included in its entirety.

The Demon and the Quantum, Second Edition. Robert J. Scully and Marlan O. Scully

ISBN 978-3-527-40983-9

Time and the Quantum: Erasing the Past and Impacting the Future

Yakir Aharonov[1,2] and M. Suhail Zubairy[3*]

The quantum eraser effect of Scully and Drühl dramatically underscores the difference between our classical conceptions of time and how quantum processes can unfold in time. Such eyebrow-raising features of time in quantum mechanics have been labeled "the fallacy of delayed choice and quantum eraser" on the one hand and described "as one of the most intriguing effects in quantum mechanics" on the other. In the present paper, we discuss how the availability or erasure of information generated in the past can affect how we interpret data in the present. The quantum eraser concept has been studied and extended in many different experiments and scenarios, for example, the entanglement quantum eraser, the kaon quantum eraser, and the use of quantum eraser entanglement to improve microscopic resolution.

The "classical" notion of time was summed up by Newton: "... absolute and mathematical time, of itself, and from its own nature, flows equally without relation to anything external." In the present article, we go beyond our classical experience by presenting counterintuitive features of time as it evolves in certain experiments in quantum mechanics. To illustrate this point, an excellent example is the delayed-choice quantum eraser, proposed by Marlan O. Scully and Kai Drühl (*1*), which was described as an idea that "shook the physics community" when it was first published in 1982 (*2*). They analyzed a photon correlation experiment designed to probe the extent to which information accessible to an observer and its erasure affects measured results. The Scully-Drühl quantum eraser idea as it was described in *Newsweek* tells the story well (*3*), and Fig. 1 is an adaptation of their account of this fascinating effect.

In his book *The Fabric of the Cosmos* (*4*), Brian Greene sums up beautifully the counterintuitive outcome of the experimental realizations of the Scully-Drühl quantum eraser (p. 149):

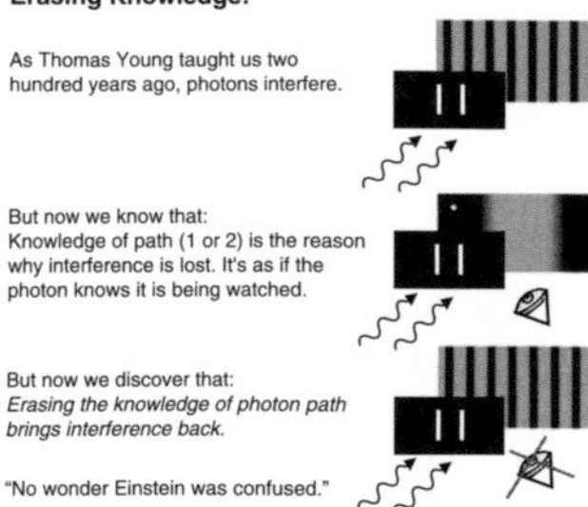

Fig. 1 Schematics for the Young's double-slit experiment. The which-path information wipes out the interference pattern. The interference pattern can be restored by erasing the which-path information.

[1]School of Physics and Astronomy, Tel Aviv University, Tel Aviv 69978, Israel. [2]Department of Physics, University of South Carolina, Columbia, SC 29208, USA. [3]Institute for Quantum Studies and Department of Physics, Texas A&M University, College Station, TX 77843, USA.

*To whom correspondence should be addressed. E-mail: zubairy@physics.tamu.edu

Enlarged Figure at end of Article.

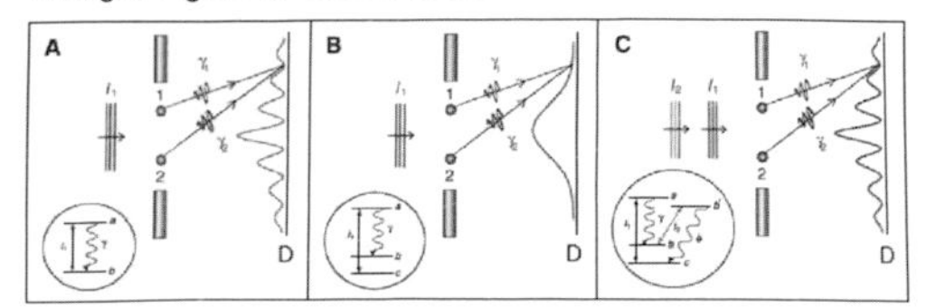

Fig. 2 Here, we consider three possible configurations of atoms that are placed at sites 1 and 2. In (**A**) we consider a two-level atom initially in the state *b*. The incident pulse l_1 excites one of the two atoms to state *a* from where it decays to state *b*, emitting a γ photon. In (**B**) the atom initially in the ground state *c* is excited by the pulse l_1 to state *a* from where it decays to state *b*. In (**C**) a fourth level is added. A pulse l_2 excites the atom to state b' after the atom has decayed to state *b*. The atom in the state b' emits a ϕ photon and ends up in state *c*.

These experiments are a magnificent affront to our conventional notions of space and time. Something that takes place long after and far away from something else nevertheless is vital to our description of that something else. By any classical-common sense-reckoning, that's, well, crazy. Of course, that's the point: classical reckoning is the wrong kind of reckoning to use in a quantum universe .... For a few days after I learned of these experiments, I remember feeling elated. I felt I'd been given a glimpse into a veiled side of reality. Common experience – mundane, ordinary, day-to-day activities – suddenly seemed part of a classical charade, hiding the true nature of our quantum world.

The world of the everyday suddenly seemed nothing but an inverted magic act, lulling its audience into believing in the usual, familiar conceptions of space and time, while the astonishing truth of quantum reality lay carefully guarded by nature's sleights of hand.

Quantum Eraser Basics

We now present a simple description of the quantum eraser that brings out the counterintuitive aspects related to time in the quantum mechanical domain. We consider the scattering of light from two atoms located at sites 1 and 2 on the screen D (Fig. 2) and analyze three different cases:

1) Resonant light impinges from the left on two-level atoms (Fig. 2A) located at sites 1 and 2. An atom excited to level *a* emits a γ photon. There are two possibilities for the atom, either it remains in the ground state *b* or it can get excited to the state *a* by the incident light and emit a γ photon. We look at the interference of these photons at the screen. Because both atoms are finally in the state *b* after the emission of photons, it is not possible to determine which atom contributed the γ photon. A large number of such experiments are carried out; i.e., any one photon will yield one count on the screen, and it takes many such photon events to build up a pattern. The resulting distribution of the detected photons exhibits an interference pattern (Fig. 2A). This is an analog of the usual Young's double-slit experiment. Instead of the usual light beams through two pin holes, we have considered scattered light from two atoms. The key to the appearance of the interference is the lack of which-path information for the photons.

2) In the case where the atoms have three levels (Fig. 2B), the drive field excites the atoms from the ground state *c* to the excited state *a*. The atom in state *a* can then emit a γ photon and end up in state *b*. Here, the photon detected on the screen leaves behind which-path information; that is, the atom responsi-

ble for contributing the γ photon is in level b, whereas the other atom remains in level c. Thus, a measurement of the internal states of the atoms provides us the which-path information and no interference is observed. That is, the state of the atom acts as an observer state. The precise mathematical description of photons γ_1 and γ_2 is the same in cases a and b. It is only the presence of the passive observer state that kills the interference.

There is an interesting connection to be made here with a statement of Richard Feynman. In his wonderful lectures on quantum mechanics for Caltech undergraduates (5), he discusses the problem of such observations rubbing out interference. He says (p. 9)

> *If an apparatus is capable of determining which hole the [photon] goes through it cannot be so delicate that it does not disturb the pattern in an essential way. No one has ever found or even thought of a way around the uncertainty principle. So we must assume that it describes a basic characteristic of nature.*

However, the loss of coherence in the present scheme does not invoke the uncertainty principle. In later work, Englert, Schwinger, Scully, and Walther came up with other such examples and in this sense have "thought of a way around the uncertainty principle" in this regard. We discuss this below.

The question, however, is whether we can erase the which-path information stored in the atom(s) and thus regain interference. If the loss of interference was caused by some kind of noise or uncertainty due to quantum fluctuations, the answer would be no. We now show that this is not the case, and the interference can be recovered. The question then is whether it is possible to wipe out the which-path information and recover the interference.

3) As shown in Fig. 2C, this can possibly be done by driving the atom by another field that takes the atom from level b to b' and, after an emission of a ϕ photon at the $b' - c$ transition, ends up in level c. Now the final state of both the atoms is c, and a measurement of internal states cannot provide us the which-path information. It would therefore seem that the interference fringes will be restored, but a careful analysis indicates that the which-path information is still available through the ϕ photon. A measurement on the ϕ photon can tell us which atom contributed the γ photon. Can we erase the which-path information contained in the ϕ photon and recover the interference fringes? Scully and Drühl considered an ingenious device based on an electrooptic shutter that can absorb the ϕ photon in such a way that the which-path information is erased (1). For the purpose of illustration, we consider a different and somewhat simplified version of such an eraser. A slightly modified version of such an eraser using a parametric process involving nonlinear crystal (instead of single atoms) was experimentally realized by Shih and co-workers in 2000 (6), which served as the motivation for Greene's presentation in (4).

However, before we proceed with discussions of Shih's experiment we should note that the erasure idea stirred up considerable controversy. Perhaps the best example is the well-written article by Mohrhoff (7). In the abstract, which we have adapted to fit the present example, he says (p. 1468)

> *a two-slit experiment ... appears to permit experimenters to choose even after each photon has made its mark on the screen, whether the photon has passed through a particular slit or has,*

in some sense, passed through both of them. Through a misleading wording the authors even appear to endorse this interpretation.

In a later paper, however, the author retracts this statement (8).

In fact, many people had a similar mind set, and it is only by carefully considering and analyzing several experiments (real and *gedanken*) that the issue is made clear.

We now turn to the particularly clear treatment of Shih and co-workers as depicted in Fig. 3. We again consider two atoms of the type shown in Fig. 2C located at sites 1 and 2. A pair of photons γ and ϕ are emitted either by the atom located at 1 or by the atom located at 2. The γ photon, as before, proceeds to the screen on the right and is detected by a detector on screen D at a location x_0. A repeat of this experiment yields an essentially random distribution of photons on the screen.

What about the appearance and disappearance of interference fringes discussed above? For this purpose, we look at the ϕ photon that proceeds to the left. We consider only those instances where the ϕ photon scattered from the atom located at 1 proceeds to the beam splitter B_1 and the ϕ photon scattered from the atom located at 2 proceeds to B_2. At either of these 50/50 beam splitters, the ϕ photon has a 50% probability of proceeding to detectors D_3 (for photon scattered from 1) and to D_4 (for photon scattered from 2). On the other hand, there is also a 50% probability that the photon will be reflected from the respective beam splitter and proceed to another 50/50 beam splitter, B. For these photons, there is an equal probability of being detected at detectors D_1 and D_2.

If the ϕ photon is detected at the detector D_3, it has necessarily come from the atom located at 1 and could not have come from the atom located at 2. Similarly, detection at D_4 means that the ϕ photon came from the atom located at 2. For such events, we can also conclude that the corresponding γ photon was also scattered from the same atom. That is, we have "which-way" information if detectors D_3 or D_4 register a count.

Returning to the quantum erasure protocol, if the ϕ photon is detected at D_1, there is an equal probability that it may have come from the atom located at 1, following the path $1B_1BD_1$, or it may have come from the atom located at 2, following the path $2B_2BD_1$. Thus, we have erased the information about which atom scattered the ϕ photon, and there is no which-path information available for the corresponding γ photon. The same can be said about the ϕ photon detected at D_2. The difference between counts in D_1 and D_2 is a phase shift such that a click at D_1 gives the fringes corresponding to $\gamma_1 + \gamma_2$, whereas a click at D_2 correlates with $\gamma_1 - \gamma_2$.

After this experiment is done a large number of times, we shall have roughly 25% of ϕ photons

Enlarged Figure at end of Article.

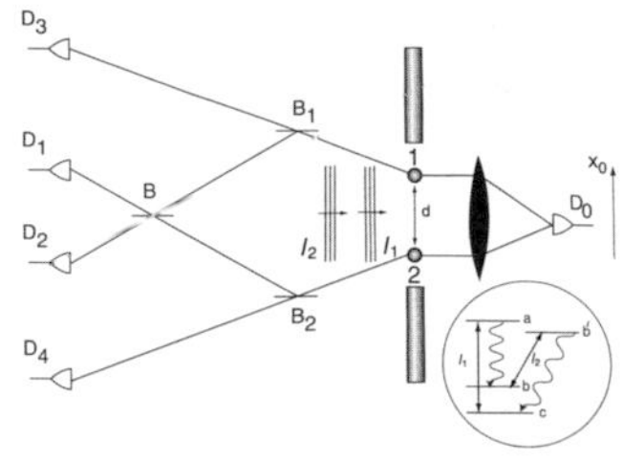

Fig. 3 Two atoms of the type shown in Fig. 2C are placed at sites 1 and 2. These atoms are excited by pulses l_1 and l_2 as in Fig. 2C such that one of the atoms emits γ and ϕ photons. We consider those events where the γ photon proceeds to the right and the ϕ photon to the left. The γ photon is collected by the detector D_0, whereas the ϕ photon is detected by D_1, D_2, D_3, or D_4 after passing through the optical setup consisting of the 50/50 beam splitters B_1, B_2, and B.

detected each at D_1, D_2, D_3, and D_4 because of the 50/50 nature of our beam splitters. The corresponding spatial distribution of γ photons will be, as mentioned above, completely random. Next we do a sorting process. We separate out all the events where the ϕ photons are detected at D_1, D_2, D_3, and D_4. For these four groups of events, we locate the positions of the detected γ photons on the screen D.

The key result is that, for the events corresponding to the detection of ϕ photons at detectors D_3 and D_4, the pattern obtained by the γ photons on the screen D is the same as we would expect if these photons had scattered from atoms at sites 1 and 2, respectively. That is, there are no interference fringes, as would be expected when we have which-path information available. On the contrary, we obtain conjugate (π phase shifted) interference fringes for those events where the ϕ photons are detected at D_1 and D_2. For this set of data, there is no which-path information available for the corresponding γ photons.

Suppose we place the ϕ photon detectors far away. Then the future measurements on these photons influence the way we think about the γ photons measured today (or yesterday!). For example, we can conclude that γ photons whose ϕ partners were successfully used to ascertain which-path information can be described as having (in the past) originated from site 1 or site 2. We can also conclude that γ photons whose ϕ partners had their which-path information erased cannot be described as having (in the past) originated from site 1 or site 2 but must be described, in the same sense, as having come from both sites. The future helps shape the story we tell of the past.

Here again the eloquent and insightful Brian Greene says it well (p. 197):

> *Notice, too, perhaps the most dazzling result of all: the three additional beam splitters and the four idler-photon detectors can only be on the other side of the laboratory or even on the other side of the universe, since nothing in our discussion depended at all on whether they receive a given idler photon before or after its signal photon partner has hit the screen. Imagine, then, that these devices are all far away, say ten light-years away, to be definite, and think about what this entails. You perform the experiment in fig 7.5b today, recording – one after another – the impact locations of a huge number of signal photons and you observe that they show no sign of interference. If someone asks you to explain the data, you might be tempted to say that because of the idler photons, which path information is available and hence each signal photon definitely went along either the left or the right path, eliminating any possibility of interference. But, as above, this would be a hasty conclusion about what happened; it would be a thoroughly premature description of the past.*

For the mathematically inclined reader we include a brief discussion (9) which sheds light on the physics using the language of modern quantum mechanics.

The Micromaser Which-Path Detector and Quantum Eraser

The Scully-Drühl quantum eraser was perhaps the earliest example of quantum entanglement interferometry and stimulated many experiments. However, another form of the quantum eraser based on cavity quantum electrodynamics and the micromaser has also stimu-

A KAON SIGNATURE

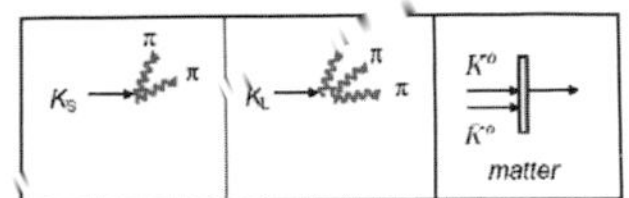

B KAON INTERFERENCE

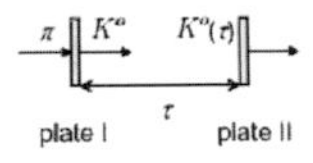

C KAON QUANTUM ERASER

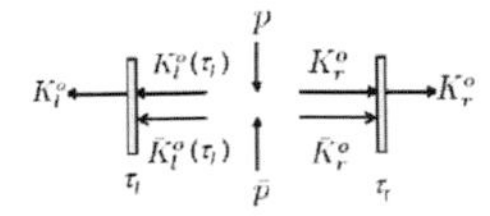

Fig. 4 (**A**) The four kaons K_S, K_L, K^0, and $\bar{K}^0$ have characteristic signatures; (the short-lived) kaon K_S decays into two π particles, whereas (the long-lived) kaon K_L decays into three π particles; the K^0 kaon (strangeness +1) mostly passes through matter, but the $\bar{K}^0$ (strangeness −1) interacts much more strongly with matter (nuclei) and is stopped. (**B**) The K^0 and $\bar{K}^0$ states are superpositions of K_S and K_L, i.e., $|K^0\rangle = (|K_S\rangle + |K_L\rangle)/\sqrt{2}$ and $|\bar{K}^0\rangle = (|K_S\rangle - |K_L\rangle)/\sqrt{2}$. Now K_S and K_L have masses m_S and m_L so that $|K^0(\tau)\rangle = (e^{-im_S\tau}|K_S\rangle + e^{-im_L\tau}|K_L\rangle)/\sqrt{2}$. Thus, if we produce K^0 particles in plate I and they propagate for a time τ to plate II then the probability for passage through plate II is $|\langle K^0|K^0(\tau)\rangle|^2$ which shows oscillations in time. (**C**) A kaon quantum eraser may be realized by noting that $p\,\bar{p}$ collisions generate the entangled states moving to the right (r) and left (l) which can be written in terms of which-way (K_S, K_L) or which-wave ($K^0, \bar{K}^0$). Quantum erasing is achieved by the left-moving kaon as the measured kaon (which will or will not show oscillations), and the right tag or ancilla kaon will serve to select the which-wave ensemble ($K^0, \bar{K}^0$) if we put in plate II and measure $K_r{}^0$. However, if we do not put in the second plate then we must describe the physics by the which-way subensemble. Thus, the entangled kaon state can be used to demonstrate quantum erasure by subensemble selection just as in the original photon case. However, if K_S or K_L propagates from I to II, the state of the kaon just before it enters II is $|K_S(\tau)\rangle = e^{-im_S\tau}|K_S\rangle$ and $|\langle K^0|K_S(\tau)\rangle|^2 = 1/2$ with a similar result for K_L. In this sense, K_S and K_L are "which-way" (short or long lived) states like photons going through slit 1 or 2, i.e., do not show oscillations. K^0 and $\bar{K}^0$, however, do show oscillation behavior and in this sense may be called "which-wave."

lated debate as well as new experiments and calculations. In particular, Englert, Schwinger (who shared the Nobel prize with Feynman and Tomonaga), Scully, and Walther showed that excited atoms passing through a microwave cavity can leave a photon in the cavity without suffering overall recoil (*10–12*).

Thus, by using the wave-like properties of the atom and placing a cavity in front of each slit, we could obtain which-way information (photon left in one cavity or the other). Furthermore, it is easy to envision ways to erase this information and regain fringes.

Again, spirited debate and decisive experiments followed. In this regard, the beautiful experiments of Dürr, Nonn, and Rempe are summarized in the following quotations taken from (*13*), where they note that the party line has it that (p. 33)

> *[If] a which-way detector is employed to determine the particle's path, the interference pattern is destroyed. This is usually explained in terms of Heisenberg's uncertainty principle.*

They further note (p. 33)

> *However, Scully et al. (10, 11) have recently*

proposed a new gedanken experiment, where the loss of the interference pattern in an atomic beam is not related to Heisenberg's position-momentum uncertainty relation.

This stirred up considerable controversy; to wit (p. 33):

Nevertheless, the gedanken experiment of Scully et al. (10, 11) was criticized by Storey et al. (14), who argued that the uncertainty relation always enforces recoil kicks sufficient to wash out the fringes. This started a controversial discussion about the following question: 'Is complementarity more fundamental than the uncertainty principle?'

They summarize their results and conclusions as follows (p. 33):

Here we report a which-way experiment in an atom interferometer in which the 'back action' of path detection on the atom's momentum is too small to explain the disappearance of the interference pattern. We attribute it instead to correlations between which-way detector and atomic motion, rather than to the uncertainty principle.

Entanglement Quantum Erasers

The preceding discussions showed how quantum eraser can be used to retrieve interference by means of tag ancilla photons $|\phi_\pm\rangle$ going with $|\gamma_\pm\rangle$ fringe and antifringe states. Garisto and Hardy (*15*) invented an interesting new class of quantum erasers, called the disentanglement eraser. These consist of at least three subsystems A, B, and T. The AB subsystem is prepared in entangled states of the type

$$|\psi_\pm\rangle = \frac{1}{\sqrt{2}} \times (|0_A, 1_B\rangle \pm |1_A, 0_B\rangle)$$

Enlarged Figure at end of Article.

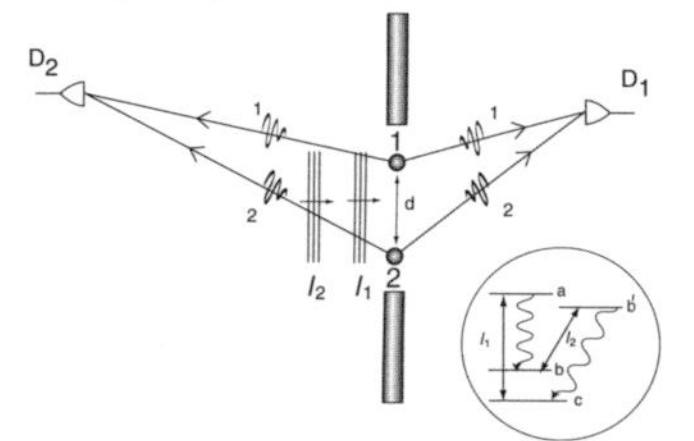

Fig. 5 Two atoms of the type shown located at sites 1 and 2 are separated by a distance d. Incident pulse sequence l_1 and l_2 leads to emission of γ and ϕ photons as in quantum eraser. The γ and ϕ photons are detected at D_1 and D_2, respectively. An intensity-intensity correlation yields resolution beyond the classical limit.

They then showed how tag states $|\phi_\pm\rangle$ can be used to remove or restore the entanglement. Thus, an outcome $|\phi_+\rangle$ for the tagged state restores the original state $|\psi_+\rangle$ for the AB subsystem, whereas the outcome $|\phi_-\rangle$ yields $|\psi_-\rangle$. Thus, a measurement of the tagging qubit restores the entangled state.

An implementation of such an eraser has been demonstrated in nuclear magnetic resonance systems (*16*). Furthermore, a cavity quantum electrodynamics–based implementation has been proposed in (*17*), which provides new insights into quantum teleportation and/or quantum dense coding.

Quantum Kaon Erasers

In a recent article (*18*), Bramon, Garbarino and Hiesmayr have extended these ideas to nuclear physics and showed that an entangled pair of neutral kaons can also display quantum erasure. In their set-up, strangeness oscillations between K^0 and $\bar{K}^0$ display oscillatory (wave-like) behavior and the alternative (which-path like) representation involving eigenstates of mass. The latter representations are called K_S and K_L because they live for about 10^{-10} and 10^{-8} s in free space. As indicated in Fig. 4, the oscillator involves a π incident on plate 1 produces a K^0 that has oscillations when expressed in terms of the K_S and K_L representation. Upon passing through the second plate, only K^0 emerges and this shows typical interference phenomena as indicated. Thus, the kaon oscillations are produced by changing the distance between the two plates. To summarize, then, with no plates we have which-way information associated with decay into two or three π particles. With the plates in place, nucleonic interactions occur, and we can observe oscillatory fringe information. Quantum eraser is achieved by using the entangled state produced by $p\,\bar{p}$ collisions.

Quantum Imaging via Quantum Eraser

Quantum interferometry using entangled photons, as in the paradigm of the quantum eraser, can be used to exceed the resolution limit of classical wave optics. The key resource needed is the ability to jointly measure and correlate the detection of two photons, as described by the intensity correlation function $G^{(2)}$. In the second order interferometry based on photon pairs, the resolution in the measurement of the distance d between the photon sources (Fig. 5) can be potentially improved by as much as an order of magnitude. In order to understand this enhanced resolution, we consider the Scully-Drühl quantum eraser configuration of Fig. 5. The atom of the type shown in Fig. 2C is first excited by a pulse l_1 of center frequency v_p and much later by a pulse l_2 at frequency v_d. A γ photon as well as a ϕ photon are emitted either by atom 1 or atom 2 that are detected by detectors D_1 and D_2. The photon-photon correlation function factorizes. The interference pattern observed by moving detector D_1 (and requiring a correlation with detector D_2) is now governed by $k_\gamma + k_\phi \approx 2k$, i.e., the effective radiation wavelength is now $\lambda/2$, leading to an immediate two-fold enhancement beyond the classical limit. In fact, Scully has shown that further improvement results from a more detailed analysis, leading to the possibility of an order of magnitude improvement of resolution (*19*).

References and Notes

1. M. O. Scully, K. Drühl, *Phys. Rev. A.* **25**, 2208 (1982).
2. S. P. Walborn, M. O. T. Cunha, S. Pádua, C. H. Monken, *Am. Sci.* **91**, 336 (2003).
3. S. Begley, *Newsweek*, 19 June 1995, p. 67.
4. B. Greene, *The Fabric of the Cosmos* (Alfred A. Knopf, New York, 2004).
5. R. P. Feynman, R. Leighton, M. Sands, *The Feynman Lectures on Physics, Vol. III* (Addison Wesley, Reading, MA, 1965).
6. Y.-H. Kim, R. Yu, S. P. Kulik, Y. Shih, M. O. Scully, *Phys. Rev. Lett.* **84**, 1 (2000).
7. U. Mohrhoff, *Am. J. Phys.* **64**, 1468 (1996).
8. U. Mohrhoff, *Am. J. Phys.* **67**, 330 (1999).
9. Mathematically we can understand the essential results of the Scully-Drühl quantum eraser by first realizing that the photon state emitted by the atoms located at sites 1 and 2 is given by

$$|\psi_0\rangle = \frac{1}{\sqrt{2}}(|\gamma_1\rangle|\phi_1\rangle + |\gamma_2\rangle|\phi_2\rangle)$$

i.e., either the photon pair γ_1, ϕ_1 is emitted by the atom located at site 1 or the pair γ_2, ϕ_2 is emitted by the atom at site 2. Thus if the ϕ photon is detected by D_3, the quantum state reduces to $|\gamma_1\rangle$. A similar result is obtained for the ϕ photon detection by the detector D_4. This is the situation when the which-path information is available and the sorted data yields no interference fringes. The physics behind the retrieval of the fringes is made clear by rewriting the state $|\psi_0\rangle$ as

$$|\psi_0\rangle = \frac{1}{\sqrt{2}}(|\gamma_+\rangle|\phi_+\rangle + |\gamma_-\rangle|\phi_-\rangle)$$

where $\gamma_\pm$ and $\phi_\pm$ are the symmetric and antisymmetric combinations.

$$|\gamma_\pm\rangle = \frac{1}{\sqrt{2}}(|\gamma_1\rangle + |\gamma_2\rangle)$$

$$|\phi_\pm\rangle = \frac{1}{\sqrt{2}}(|\phi_1\rangle + |\phi_2\rangle)$$

The state of the ϕ photon after passage through the beam splitter B is either $|\phi_+\rangle$ or $|\phi_-\rangle$. Thus, a click at detectors D_1 or D_2, reduces the state of the γ photon to $|\gamma_+\rangle$ or $|\gamma_-\rangle$, respectively, leading to a retrieval of the interference fringes.

10. M. O. Scully, B.-G. Englert, H. Walther, *Nature* **351**, 111 (1991).
11. B.-G. Englert, J. Schwinger, M. O. Scully, *Found. Phys.* **18**, 1045 (1988).
12. M. O. Scully, M. S. Zubairy, *Quantum Optics* (Cambridge, London, 1997).
13. S. Durr, T. Nonn, G. Rempe, *Nature* **395**, 33 (1998).
14. P. Storey, S. Tan, M. Collett, D. Walls, *Nature* **367**, 626 (1994).
15. R. Garisto, L. Hardy, *Phys. Rev. A.* **60**, 827 (1999).
16. G. Teklemariam, E. M. Fortunato, M. A. Pravia, T. F. Havel, D. G. Cory, *Phys. Rev. Lett.* **86**, 5845 (2001).
17. M. S. Zubairy, G. S. Agarwal, M. O. Scully, *Phys. Rev. A.* **70**, 012316 (2004).
18. A. Bramon, G. Garbarino, B. C. Hiesmayr, *Phys. Rev. Lett.* **92**, 020405 (2004).
19. M. O. Scully, unpublished results.

20. We thank E. Fry, A. Muthukrishnan, R. Ooi, and A. Patnaik for their help in the preparation of this manuscript. We also gratefully acknowledge support from U.S. Air Force Office of Scientific Research, Defense Advanced Research Projects Agency, and Texas A&M University's Telecommunication and Informatics Task Force initiative.

10.1126/science.1107787

E9.2 Quantum Eraser Analysis

This section is a bit more advanced than most aspects of the end notes and is intended for more advanced readers.

Let us consider first the atomic beam of Fig. 9.2 without the laser/maser cavities. After passing through the double slits, the wave function for the atom is

$$\Psi(\mathbf{r}) = \frac{1}{\sqrt{2}}[\psi_1(\mathbf{r}) + \psi_2(\mathbf{r})], \tag{1}$$

where $\mathbf{r}$ is the atomic center-of-mass coordinate. Hence the probability density for particles on the screen at $\mathbf{r} = \mathbf{R}$ is given by the squared modulus of $\Psi(\mathbf{R})$,

$$P(\mathbf{R}) = \frac{1}{2}[|\psi_1|^2 + |\psi_2|^2 + (\psi_1^*\psi_2 + \psi_2^*\psi_1)]. \tag{2}$$

Next, consider the micromaser cavities in the two paths. After passing through the cavities, the state of the correlated atomic beam and cavity field is now

$$\Psi(\mathbf{r}) = \frac{1}{\sqrt{2}}[\psi_1(\mathbf{r})|1_1 0_2\rangle + \psi_2(\mathbf{r})|0_1 1_2\rangle], \tag{3}$$

where $|1_1 0_2\rangle$ denotes the state in which there is one photon in cavity 1 and none in cavity 2, etc. The probability density at the screen is now given by

$$P(\mathbf{R}) = \frac{1}{2}[|\psi_1|^2 + |\psi_2|^2 + (\psi_1^*\psi_2\langle 1_1 0_2|0_1 1_2\rangle + \psi_2^*\psi_1\langle 0_1 1_2|1_1 0_2\rangle)]. \tag{4}$$

But because of the vanishing inner product $\langle 1_1 0_2 | 0_1 1_2 \rangle$, the interference terms disappear, so that we now have

$$P(\mathbf{R}) = \frac{1}{2}[|\psi_1|^2 + |\psi_2|^2]. \tag{5}$$

Thus, we see that it is the system-detector correlations, which account for the dramatic effects of the measuring apparatus on the system of interest. It is no surprise that coherence is destroyed as soon as one has which-way information, but here no uncontrollable scattering events were involved in destroying the interference (wave-like) behavior.

One then wonders whether it might not be possible to retrieve the coherent interference cross-terms by removing ('erasing') the which-way information contained in the detectors. The affirmative answer to this question is given mathematically as follows. Extending the description to include the detector, which is initially in its ground state d, we have

$$\Psi(\mathbf{r}) = \frac{1}{\sqrt{2}}[\psi_1(\mathbf{r})|1_1 0_2\rangle + \psi_2(\mathbf{r})|0_1 1_2\rangle]|d\rangle. \tag{6}$$

After absorbing a photon, the detector would be found in the excited state e. We now introduce symmetric, ψ_+, and anti-symmetric, ψ_-, atomic states defined as

$$\psi_\pm(\mathbf{r}) = \frac{1}{\sqrt{2}}[\psi_1(\mathbf{r}) \pm \psi_2(\mathbf{r})]. \tag{7}$$

We also introduce symmetric, $|+\rangle$, and anti-symmetric, $|-\rangle$, states of the radiation fields contained in the which-way cavities,

$$|\pm\rangle = \frac{1}{\sqrt{2}}[|1_1 0_2\rangle \pm |0_1 1_2\rangle]. \tag{8}$$

The initial atom-beam/microwave-cavity/detector system now is given by

$$\Psi(\mathbf{r}) = \frac{1}{\sqrt{2}}[\psi_+(\mathbf{r})|+\rangle + \psi_-(\mathbf{r})|-\rangle]|d\rangle. \tag{9}$$

Next we consider the interaction between the radiation field existing in the cavity and the detector. We envision the detector to consist of an atom with a lower state d and an excited state e. The interaction between field and detector depends on symmetric combinations of the field variables, so only the symmetric state will couple to the fields.

We find that the action of the eraser detector produces the state

$$\Psi(\mathrm{r}) = \frac{1}{\sqrt{2}}[\psi_+(\mathrm{r})|0_1 0_2\rangle|e\rangle + \psi_-(\mathrm{r})|-\rangle|d\rangle]. \tag{10}$$

because the symmetric interaction between the atom and the field couples only to the symmetric radiation state $|+\rangle$.

The probability $P_e(\mathbf{R})$ for finding both the detector excited and the atom at **R** on the screen is now

$$\begin{aligned} P_e(\mathbf{R}) &= |\psi_+(\mathbf{R})|^2 \\ &= \frac{1}{2}[|\psi_1(\mathbf{R})|^2 + |\psi_2(\mathbf{R})|^2] + \mathrm{Re}[\psi_1^*(\mathbf{R})\psi_2(\mathbf{R})], \end{aligned} \tag{11}$$

which exhibits fringes.

To summarize, after an atom has passed from the oven to the screen, passing through micromasers, we record an event somewhere on the screen. Then we return to the which-way micromasers, open the shutters and allow the absorption of the microwave photon. When we observe a photon count, we know that erasure has been completed, and this is counted as a 'yes'-event. Then we wait for another atom to pass through and record an event on the screen and then turn to the micromaser cavities. This time suppose that, upon opening the shutter, we observe no photon count in the quantum eraser detector. Then we count the atom as a 'no'-event. We repeat the above many times. Eventually, the 'yes'-events will build up fringes, and the 'no'-events antifringes. Finally we note that the fringes and antifringes will cancel (i.e., they will add and wipe out the interference pattern) if Rob does not correlate them to the state of the eraser-detector.

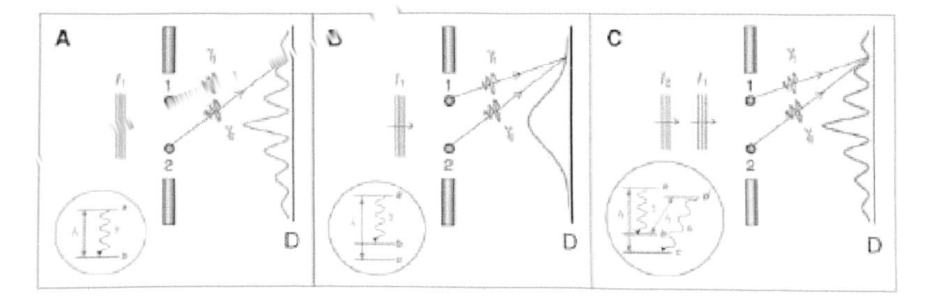

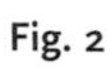
Fig. 2

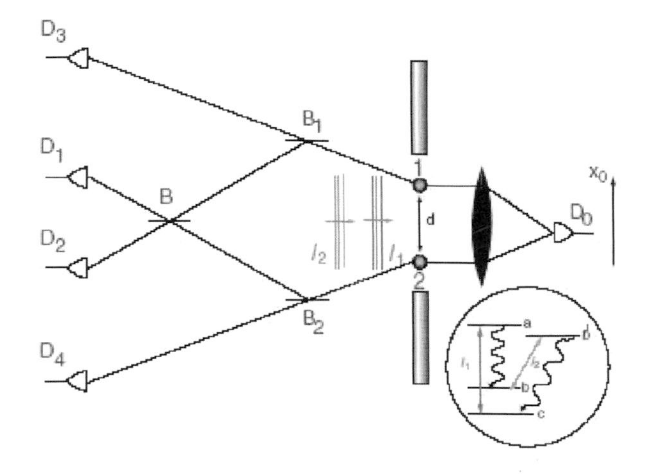

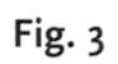
Fig. 3

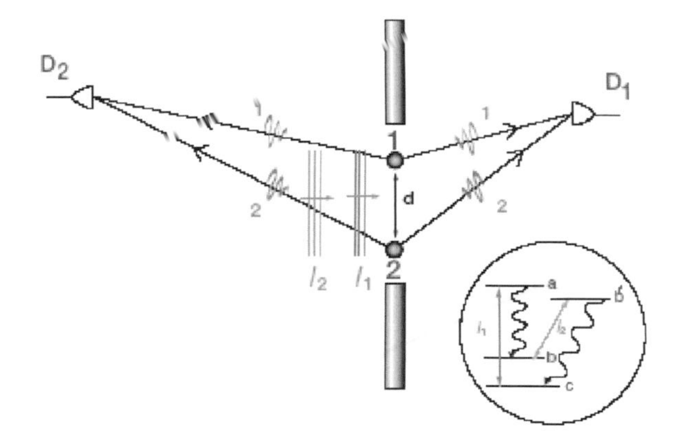

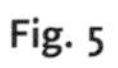
Fig. 5

Endnotes 10

Three articles and/or parts of articles on entanglement and quantum eraser are included:

1. "An End to Uncertainty" (New Scientist pp 24–28, Vol. 161, No. 2176 (1999)) by Mark Buchanan.
2. Excepts from: "Restoration of interference and the fallacy of delayed choice: Concerning an experiment proposed by Englert, Scully, and Walther" (American Journal of Physics, 64(12), 1468–1475 (1996)) and "Objectivity, retrocausation and the experiment of Englert, Scully, and Walther" (American Journal of Physics 67, 330 (1999)) both by Ulrich Mohrhoff.
3. Excerpts from: "Quantum Erasure in Double-Slit Interferometers with Which-Way Detectors" (American Journal of Physics 67, 325 (1999)) by Berthold-Georg Englert, Marlan O. Scully, and Herbert Walther.

E10.1 Summary of Experiment Showing that Certainty Beats Uncertainty

The uncertainty relation – quantum entanglement debate is explained and the experiments of Rempe et al., resolving the issue is very well presented in the following article by Buchanan.

The Demon and the Quantum, Second Edition. Robert J. Scully and Marlan O. Scully

ISBN 978-3-527-40983-9

An end to uncertainty

Wave goodbye to the uncertainty principle —you don't need it any more. Say hello to quantum entanglement, says *Mark Buchanan*

EINSTEIN versus Bohr: one of the most famous bouts in science. For the years spanning the late 1920s and early 1930s, these two fought over the future of physics. Albert Einstein could not accept the outrageous randomness and unknowability of quantum mechanics, so he attacked the theory by devising a series of ingenious thought experiments. But whenever he seemed to have nailed an inconsistency at the core of quantum theory, Niels Bohr proved him wrong. Despite all its unpalatable ingredients, quantum mechanics won the day.

Bohr invariably demolished Einstein by using Werner Heisenberg's uncertainty principle. Measure the position of an electron or any other quantum particle, and, according to Heisenberg, you will disturb its momentum. Measure its momentum, and you will disturb its position. So you can never know both the momentum and the position of a particle at once. Ever since Bohr used this idea to win his legendary victory, the uncertainty principle has stood as the conceptual heart of quantum theory.

Bohr and Einstein had to devise imaginary experiments to prove their theories, because the technology to do the real experiments just didn't exist. That's changed. On a table strewn with delicate lasers, Gerhard Rempe and his colleagues at the University of Konstanz in Germany have brought to life one of the most famous experiments that the giants of quantum theory argued over. And they are using it to turn history on its head.

Quantum mechanics is still standing tall, but it now appears that Niels Bohr won his famous victory with faulty arguments. He inadvertently misled Einstein, and for 70 years most physicists have misunderstood the most important physical theory there is. They have been labouring under the delusion that what makes quantum theory so weird is its inherent uncertainty, or fuzziness, but in fact another feature of the quantum world, a phenomenon called entanglement, is at the root of it all. So what has ended these decades of delusion?

The experiment is, in its logic, astonishingly simple. It is the well-known two-slit experiment, which shows up one of the quantum world's deepest mysteries: how something can be both a wave and a particle.

The idea is to send a beam of particles towards a barrier with two slits in it and see where they hit a detecting screen beyond (see Diagram, p 27). According to quantum mechanics, the result at the screen is an interference pattern, a set of parallel dark and bright bands. This shows that the beams going through the slits act like waves, which either reinforce or cancel each other out depending on where they meet. The same pattern is built up, particle by particle, even if the beam is so weak that only one particle goes through per hour, say.

This is what so unnerved Einstein: how can a single particle interfere with itself? How does it know that both slits are open, and cooperate in forming the interference pattern? Quantum mechanics says that it must somehow split into two ghosts of a particle, one going through each slit, which interfere with each other on the other side.

Why not test this strange idea by simply looking to see which way the particle goes? Shine some light near the slits, and you will see a few photons bounce off the particle as it goes through one hole or the other, proving that the particle doesn't go through both holes – but you'll still see an interference pattern. Surely this should show that the idea of a particle interfering with itself is nonsense?

But it doesn't. Using the uncertainty principle, Bohr and Heisenberg destroyed any hope that this ploy could work. To be able to tell which slit the particle goes though, the argument goes, you must fix its position to a precision better than the distance between the slits.

Split personality

Heisenberg's uncertainty principle demands that if you pin down the particle's position so precisely, you must increase the uncertainty in its momentum. Bohr said that this happens because the photons deliver random, uncontrollable momentum kicks as they bounce off the particle. This disturbance changes the position where the particle hits the screen by a distance that is about as large as the spacing between the interference bands, so the pattern inevitably gets smeared away. In other words, if you look to see which way the particle goes, there's no interference, so Einstein's hoped-for contradiction evaporates.

Physicists managed to do this thought experiment for real only in the early 1990s, but the results were exactly as Bohr and Heisenberg said they would be. If you look to see which way the particles go they stop acting like waves, and the pattern on the screen is a big blob, not an interference pattern.

For half a century, physicists have memorised, repeated and regurgitated this story of how the uncertainty principle acts as the invincible defender of quantum theory. Learning it is virtually a rite of initiation for aspiring physicists. Not surprising, then, that when Rempe and his colleagues reported the results of their experiment last September, there was consternation in the ranks. Bohr's reasoning, their results prove, is based on a fallacy.

The essence of the new experiment was proposed in 1991 by Marlan Scully, Berthold-Georg Englert and Herbert Walther of the Max Planck Institute for Quantum Optics in Garching, Germany. A two-slit experiment works with any kind of quantum particle but they suggested that atoms might offer an advantage. An atom has a variety of different internal states: a lowest energy ground state and a series of higher energy or "excited" states. And these different states, they reckoned, could be used to record the atom's path.

"Much of our experiment is based on that proposal," says Rempe. In the 1980s, physicists devised ways to cool atoms to within a hair's breadth of absolute zero using laser light. "Scully and his colleagues came up with the idea because they could use cold atoms," says Rempe. The point about cold atoms is that they have long wavelengths, which makes their interference patterns relatively easy to observe.

Still, no one could make the experiment work until last year, when Rempe and his colleagues managed it with a few clever tricks. They didn't actually send atoms through slits in a solid barrier, but instead split a beam of cold rubidium atoms using thin barriers of pure laser light (see Diagram, p 28). The beams overlap, but travel along slightly different paths, A and B. As in the classic two-slit experiment, the two beams then combine to create an interference pattern.

But then Rempe and his colleagues looked to see which path the atoms followed. The atoms going down path A weren't interfered with, but those on path B were tweaked into a higher energy state by a pulse of microwaves. So the atoms, in their internal states, kept a record of which way they had gone.

The payoff is impressive. The microwaves have hardly any momentum of their own, so they can cause little change to the atom's momentum – certainly not enough to smear away the interference pattern.

Yet the quantum world's wave-particle balancing act still works. With the microwaves turned off, the interference fringes appear. Turn them on, so that you can tell which way the atoms went, and the fringes suddenly vanish. "Everyone believes that when an interference

'For half a century physicists have memorised, repeated and regurgitated the story of how the uncertainty principle acts as the invincible defender of quantum theory. So Rempe's experiment caused consternation'

pattern is lost, it happens because a measuring device delivers random kicks to the particles. But there are no random kicks in our experiment," says Rempe. At least, none worth mentioning. Rempe estimates that, at worst, the microwaves deliver momentum kicks ten thousand times too small to destroy the interference fringes. The uncertainty principle isn't proved wrong, because in this setup the measurement of position is very imprecise, but it can't explain the results.

So what's going on? Is the central story of quantum theory just that – a story? Or is this one experiment merely an unimportant curiosity? At the University of Cambridge, physicist Yu Shi is trying to find out. Motivated by Rempe's experiment, he has taken another look at the early thought experiments in which Bohr "defeated" Einstein. And he has come to be less than impressed by Bohr's analyses.

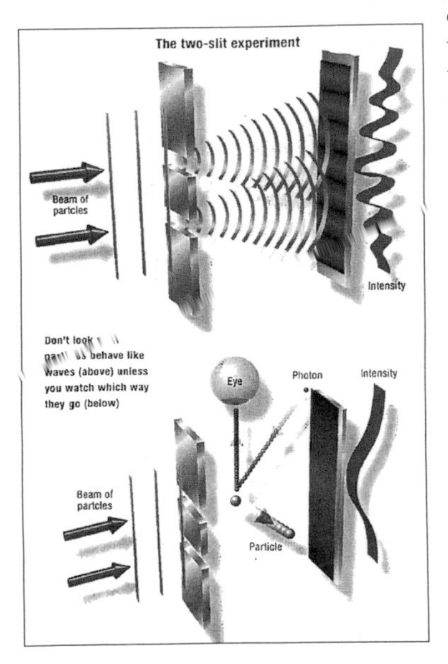

Each of these thought experiments was designed to portray a particular case in which the quantum world refuses to reveal both its wave-like and its particle-like faces at the same time. And in each case, Shi points out, Bohr discussed the physics using only the simple Planck and de Broglie relations. These are the rudimentary equations that connect a particle's momentum and energy to its wavelength and frequency.

So Shi has reanalysed the thought experiments using the rigorous equations of quantum theory, which give the fullest description possible of a quantum particle. And he has found that despite everything Bohr said, the uncertainty principle never has anything to do with destroying the interference. "People think that Bohr was right, and Einstein was wrong," he says, "but this is far from the truth. Bohr's idea that a momentum kick destroys the interference is wrong."

Shi's point is that although momentum kicks seem to explain the classic two-slit experiment, it is just a happy coincidence of numbers. There is a far deeper mechanism at work: it is the getting of path information itself that spoils the interference, says Shi. Forget all vague ideas of uncertainty, and look instead to the far more precise notion of "quantum entanglement".

Inextricably linked

Ordinarily, we regard separate objects as independent of one another. They live on their own terms, and anything tying

'This is the essence of entanglement: the interaction pairs up each atomic ghost with a corresponding photon ghost'

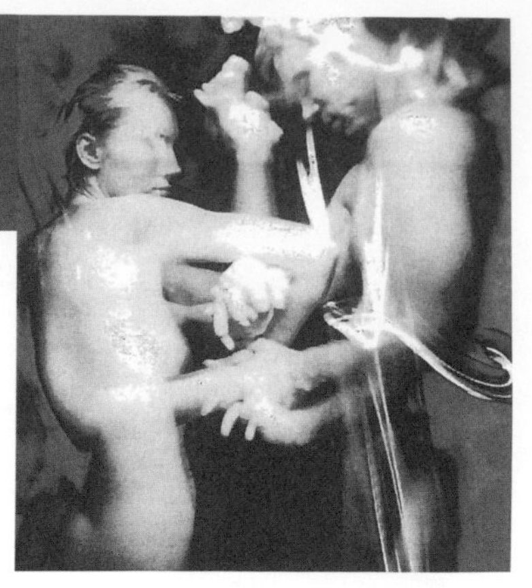

them together has to be forged by some tangible physical mechanism. Not so in the quantum world. If a particle interacts with some object – another particle, perhaps – then the two can become inextricably linked, or entangled (see "Beyond reality", *New Scientist*, 14 March, 1998, p 26). In a sense, they simply cease to be independent things, and one can only describe them in relation to each other.

What does this do to a particle's ability to show wave-like behaviour? By itself, an atom can act as a wave. In a two-slit device, however, it effectively splits its own existence, and goes through both slits. If these two ghosts of the atom move along their paths without running into anything, then they recombine and interfere at the wall.

But suppose you send a photon towards one of the slits. If an atom were there, the photon would simply bounce off and record the atom's position. But because the atom's identity is already split between the two paths, it makes the photon split too. A ghost of the photon bounces off the ghost of the atom at that slit, and a second photon ghost carries straight on. This is the essence of entanglement – the interaction pairs up each atom ghost with a corresponding photon ghost. Linked with their photon parasites, the two atom ghosts are mismatched, so the interference vanishes.

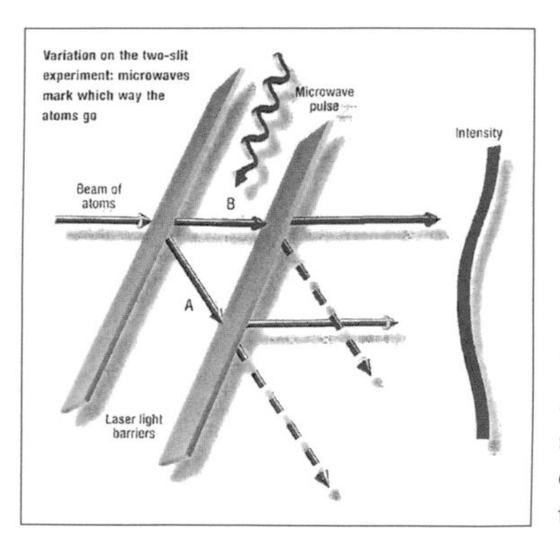

Variation on the two-slit experiment: microwaves mark which way the atoms go

"Loss of interference is always due to entanglement," says Shi, who sees in it the true origin of quantum weirdness. Quantum particles can split into ghosts that can move on many paths at once, and when they come back together we see wave-like behaviour and interference patterns. But reach into the quantum world, and you will inevitably attach disruptive partners to the quantum ghosts, partners that will spoil the reunion and make the ghosts act as if they were true particles.

Awe of Bohr

Why has it taken physicists so long to appreciate this? Shi suspects that is was simple confusion. In 1935 Einstein recognised that if two particles were entangled, doing something to one could immediately affect the other, even at a great distance. As a result, Einstein doubted that entanglement could be real. But since then, experiments have provided strong evidence that this "nonlocal" linking of distinct parts of the world really happens ("Why God plays dice", *New Scientist*, 22 August 1998, p 26). Because this effect is so shocking, physicists have dismissed entanglement as a nonlocal effect, and missed its role in the simple experiments.

Overlooking entanglement, physicists have instead taken Bohr's word as gospel. And most of them still share Bohr's exaggerated opinion of Heisenberg's uncertainty principle. "I think people give too high a position to it," says Shi. He now sees the idea as little more than a conceptual half-way house – something that allows physicists to talk and think about quantum particles as if they were vaguely classical, and to "explain" their strange ways with a fuzzy form of ordinary mechanics.

So for 70 years physicists have been explaining some of the most elementary of quantum happenings in terms of a principle that, although true, turns out to be irrelevant. But what does this mean for physics? For the day-to-day business of making quantum calculations, probably nothing. But for physicists' understanding of why they believe what they believe, the implications are unsettling.

Physicists like to pride themselves on their questioning minds, and their fierce intellectual independence. But it seems that the uncertainty principle, and the nebulous picture of the quantum world it supports, owe much of their status to little more than Bohr's exalted reputation as a physicist. "Bohr was notorious for being obscure in his writings," says Maria Beller, a historian and philosopher of science at the Hebrew University of Jerusalem. And yet many physicists, she points out, have referred to that obscurity as reflecting the "depth and subtlety" of his thought, even if they couldn't really work out what he was saying. She cites the German physicist Carl von Weizsäcker as a typical victim of what she calls "the overpowering, almost disabling, impact of Bohr's authority".

She tells the story of how after once visiting Bohr to discuss physics, von Weizsäcker found himself wondering what Bohr had meant. "I tortured myself," he recalled, "on endless solitary walks." And yet von Weizsäcker never considered the possibility that Bohr might be wrong. "Quite incredibly," Beller says, "he wondered what must one assume and in what way must one argue in order to render Bohr right".

But maybe Bohr was so obscure that he couldn't help but be right. He liked to talk about "the profound truths" for which "the opposite is also the truth". And he also admitted that "every sentence I utter must be understood not as an affirmation, but as a question". Perhaps physicists are beginning to catch on. □

Mark Buchanan is a freelance science writer

Further reading: "Origin of quantum mechanical complementarity probed by a 'which-way' experiment in an atom interferometer" by S. Dürr, T. Nonn & G. Rempe, *Nature* vol. 395, p 33

E10.2 Quantum Eraser Debate with Mohrhoff

In section E7.1, we noted that vigorous debate is essential to scientific progress. However, unlike many fields, the points of disagreement are usually sorted out and resolved in short order. Everyone profits as the issues sharpen, and our understanding deepens.

The next four pages summarize a debate between Ulrich Mohrhoff and ESW (Englert, Scully and Walther), on quantum eraser. Mohrhoff is obviously a bright and strong-minded individual, and his arguments are worth studying.

In the process of resolving our differences, we (ESW) were also able to clarify how it is that once one understands quantum eraser, and to underscore the fact that the same line of thought resolves and clarifies the EPR "paradox" as per the earlier paper of Cantrell and Scully.

A short summary of two papers by Mohrhoff on quantum eraser and the ESW reply follow:

Restoration of interference and the fallacy of delayed choice: Concerning an experiment proposed by Englert, Scully, and Walther

Ulrich Mohrhoff
Sri Aurobindo Ashram, Pondicherry 605002, India

(Received 28 November 1995; accepted 4 March 1996)

A two-slit experiment with atoms, proposed by Englert, Scully, and Walther, appears to permit experimenters to choose, after each atom has made its mark on the screen, whether the atom has passed through a particular slit or has, in some sense, passed through both of them. Through a misleading wording these authors even appear to endorse this interpretation. In actual fact, this choice exists only until the atom hits the screen. The said experiment thus is a "delayed-choice" experiment only in the semantically contingent sense of Wheeler.

I. Introduction

Englert, Scully and Walther (ESW) have proposed a two-slit experiment with atoms in which (so they appear to claim) the experimenter can decide, long *after* an atom has made its mark on the screen, whether that atom has passed through a particular slit or has, in a sense, passed simultaneously through both of them. If this were true, the ESW experiment would provide much stronger evidence of a delayed choice concerning the observable to be measured than the so-called delayed-choice experiments discussed by Wheeler. To the already sufficiently baffling features of quantum mechanics (the ability of particles to travel simultaneously along different routes; the nonseparability, under certain conditions, of spatially well-separated quantum systems) we would have to add what might be the most baffling of all: the possibility of causally determining the past. (Weeler's "delayed-choice" experiments make this conclusion far less inescapable, as we will see.)

The purpose of this paper is to show that, in fact, the experimenter does not have the above-mentioned choice. (Nor, as will become clear, is this actually asserted by

ESW, notwithstanding some misleading statement by these authors.) We are spared the conclusion that action into the past is possible, unless we go along with Wheeler's unconventional interpretation of the quantum formalism, where delayed choice has its semantic niche. The ESW experiment qualifies as a delay-choice experiment only in the restricted sense in which "Wheeler's experiments qualify. And even in this restricted sense, the possibility of choosing between alternative "histories" exists only *until* the atom makes its mark on the screen.

Objectivity, retrocausation, and the experiment of Englert, Scully, and Walther

Ulrich Mohrhoff
Sri Aurobindo Ashram, Pondicherry 605002, India

(Received 5 August 1998; accepted 17 September 1998)

In a recent contribution to this journal [Am. J. Phys. **64**, 1468–1475 (1996)] I wrongly asserted that retrocausation in the Englert, Scully, and Walther (ESW) experiment (a double-slit interference experiment with atoms) can occur only until the atom arrives at the screen. In their response, Englert, Scully, and Walther [preceding paper] point out my fallacy but give an incomplete analysis of its origin. In this paper I trace this fallacy to a deep-seated preconception about time and reality. I show that among the two possible realistic interpretations of standard quantum mechanics, the reality-of-states view and the reality-of-phenomena view, only the latter is viable. It follows that retrocausation is a necessary feature of any realistic account of the ESW experiment based on standard quantum mechanics. Finally I eludicate the sense in which the spatial properties of quantum systems are objective, and show that they are extrinsic rather than intrinsic.

I. Introduction

In a recent article I analyzed the thought experiment of Englert, Scully, and Walther (ESW) from two "metaphysical" perspectives, the reality-of-states view and the reality-of-phenomena view. In that article I arrived at a wrong conclusion, for which I wish to express my sincere apologies to the readers of this journal. I compounded my mistake by attributing my views to Englert, Scully, and Walther. My apologies also to these authors! It ought to be mentioned, however, that I was argued into misrepresenting their views by the anonymous referee of my article.

Quantum erasure in double-slit interferometers with which-way detectors

Berthold-Georg Englert
Max-Planck-Institut für Quantenoptik, Hans-Kopfermann-Strasse 1, 85748 Garching, Germany and Atominstitut der Österreichischen Universitäten, Stadionallee 2, 1020 Wien, Austria

Marlan O. Scully
Max-Planck-Institut für Quantenoptik, Hans-Kopfermann-Strasse 1, 85748 Garching, Germany and Department of Physics, Texas A&M University, College Station, Texas 77843

Herbert Walther
Max-Planck-Institut für Quantenoptik, Hans-Kopfermann-Strasse 1, 85748 Garching, Germany and Sektion Physik, Universität München, Am Coulombwall 1, 85748 Garching, Germany

(Received 21 July 1997; accepted 17 September 1998)

Recently, Mohrhoff [Am. J. Phys. **64**, 1468–1475 (1996)] has analyzed a thought experiment of ours [Nature (London) **351**, 111–116 (1991)] where a double-slit interferometer for atoms is supplemented by a pair of which-way detectors. Owing to the quantum nature of these detectors, the experimenter can choose between acquiring which-way knowledge and observing an interference pattern. The latter option makes use of a procedure called "quantum erasure." Mohrhoff (along with other bright colleagues who have made similar statements) claims erroneously that the experimenter has to make this choice before the atom hits the screen. We readdress this issue here and demonstrate that our original assertion was correct: The experimenter can choose between which-way knowledge and quantum erasure at any time, even after the atom has left its mark on the screen.

I. Introduction

Some conclusions reached by Mohrhoff in his recent analysis1 of the thought experiment that we had introduced earlier publications are erroneous. The errors are not in Mohrhoff's mathematical equations but in the physical interpretation that he attaches to some of them. It is the objective of this paper to set the record straight. In Sec. II we recall the essential features of our thought experiment: making which-way information available in a double-slit setup and then (1) using it to sort the atoms into subensembles ("first slit or second slit?"); or (2) performing quantum erasure, by which subensembles are recognized that exhibit interference fringes. This is followed by a discussion of Mohrhoff's point of view in Sec. III, and we identify the trap into which he fell. We conclude with a few additional remarks in Sec. IV and deal with related issues in the Appendix.

IV. Concluding Remarks

It appears that Mohrhoff is led astray by regarding the state reductions (6) and (8) as physical processes, rather than accepting that they are nothing but mental processes. This point of view necessarily requires that the original state vector $|\Psi\rangle(x)$ as well as the reduced ones are regarded as real physical objects, rather than as the bookkeeping devices that they are in the first place. We recall: The state vector $|\Psi\rangle(x)$ serves the sole purpose of summarizing concisely our knowledge about the entangled atom-and-photon system; in conjunction with the known dynamics, it enables us to make correct predictions about the statistical properties of future measurements. And a state reduction must be performed whenever we wish to account for newly acquired information about the system. This *minimalistic interpretation* of state vectors and their reduction is common to all interpretations; it is forced upon us by the abundance of empirical facts that how that quantum mechanics works. Of course, one might try to go beyond the minimalistic interpretation and give additional ontological meaning to $|\Psi\rangle(x)$, thereby accommodating some philosophical preconceptions or other personal biases. In doing so, one should however remember van Kampen's caveat: Whoever endows the state vector with more meaning than is needed for computing observable phenomena is responsible for the consequences). We think that Mohrhoff's fallacy is an example of the kind of trap that van Kampen is warning against.

Acknowledgments

I wish to thank the many generous people who have improved this book by their many criticisms and suggestions. The following is an incomplete list.

John Alvis	Professor of English, University of Dallas
Harold Bailey	Mechanical Engineer
Bradley, Walter	Distinguished Professor, Baylor University
Kimberly Chapin	Administrative Assistant, TAMU*
Gordon Chen	Professor of Mathematics, TAMU
Siu Chin	Professor of Physics, TAMU
William Bedford Clark	Professor of English, TAMU
Leon Cohen	Professor of Physics, City Univ. N.Y.
Michael Cone	Research Assistant, TAMU
David Depatie	Physicist
George Ellis	Professor, University of Cape Town
Noam Erez	Research Physicist, Weizmann Institute
Francis Everitt	Professor of Physics, Stanford
Bob Kaita	Physicist, Princeton Plasma Physics Lab
Michael Fisher	Chief Editor, Harvard University Press
Ken Ford	Physicist–Writer
Chris Fuchs	Physicist, Lucent Research Laboratory
Owen Gingerich	Professor of Astronomy, Harvard University
Roy Glauber	Nobel Prize Winner, Quantum Optics
Alexander Grossmann	Chief Editor, Wiley-VCH
Dudley Herschbach	Nobel Prize Winner, Chemistry
Clayton Holle	Administrative Assistant, TAMU
Andrew Jordan	Assistant Professor, University of Rochester
Marian Jordan	Editorial Advisor

* TAMU = Texas A&M University

The Demon and the Quantum, Second Edition. Robert J. Scully and Marlan O. Scully

ISBN 978-3-527-40983-9

Peter Keefe	Lawyer–Physicist
Barnabas Kim	Senior Research Assistant, TAMU
Donald Kobe	Professor, University of North Texas
Norbert Kroó	President Hungarian Acad. of Sciences
Becky Lincoln	Editorial Advisor
Ashok Muthukrishnan	Assistant Professor, Swarthmore College
Anil Patnaik	Research Physicist, TAMU
Yuri Rostovtsev	Research Associate Professor, TAMU
Zoe Sariyanni	Research Assistant, TAMU
Steve Scully	Electrical and Computer Engineer
Judy Scully	Advisor, Consultant, and Critic
Szymon Suckewer	Professor, Princeton University
Charles H. Townes	Nobel Prize, Invention of the Maser
Anja Tschörtner	Physics Editor, Wiley-VCH
Stan Wakefield	Editorial Advisor and Consultant
Ray Weymann	Astronomer
Reese Woodling	Past Pres., Internat. Brangus Assoc./Rancher
Jerry Youmans	Medical Doctor/Editorial Advisor
Kristen Youmans	Theologian
M. Suhail Zubairy	Professor of Physics, TAMU

Glossary of Terms

Absolute zero This is the coldest attainable temperature, corresponding to 0 K on the Kelvin scale, −459.67 °F on the Fahrenheit scale, and −273.15 °C on the Celsius scale. It is completely devoid of heat or thermodynamic energy.

Arrow of time Physical processes at the microscopic level are time-symmetric, yet at the macroscopic level there is an obvious direction (or arrow) of time. The statistical arrow of time posits that it is the increase of entropy that breaks the symmetry.

Bar magnets This refers to the shape of the magnets. A magnet's shape determines the effect it will have on the observed object. Two bar magnets, because they have the same shape, will deflect (repel) particles with a magnetic moment to align them with either their north pole or their south pole. See Figs. 6.1 and 6.2.

Beam combiner A device designed to combine two separate beams of light or atoms. See Figs. 7.3 and 7.5.

Big bang A cosmological theory holding that the universe (space and time) originated approximately 20 billion years ago from the violent explosion of a very small agglomeration of matter of extremely high density and temperature.

Black body An ideal black substance that absorbs all and reflects none of the radiation falling on it. Since a black body is a perfect absorber of radiant energy, by the laws of thermodynamics it must also be a perfect emitter of radiation. The distribution according to wavelength of the radiant energy of a black-body radiator depends only on the absolute temperature of the black body. In order to explain the spectral distribution of black-body radiation, Max Planck developed the quantum theory in 1901.

The Demon and the Quantum, Second Edition. Robert J. Scully and Marlan O. Scully

ISBN 978-3-527-40983-9

Boltzmann constant The physical constant (denoted by k or k_B) relating temperature to energy. It is named after the Austrian physicist Ludwig Boltzmann, who made important contributions to the theory of statistical mechanics, in which this constant plays a crucial role. Its experimentally determined value (in SI units) is $k = 1.380\,6505 \times 10^{-23}$ J/K.

Calculus The branch of mathematics that deals with limits, and the differentiation and integration of functions.

Caloric theory The long-held assumption that every flammable substance contained a fluid called caloric; the caloric was consumed during burning. Thus flammable fluids disappeared completely when burned, while logs became greatly diminished piles of ash. This theory was undone by modern discoveries in chemistry and thermodynamics.

Carnot engine A theoretical engine that operates with maximum efficiency by operating ultra-slowly and via appropriate temperature control. See Fig. 3.2.

Classical physics The school of physics set forth by Newton, characterized by deterministic physical laws predicting forces and motion. It was later in conflict with, and would be extended by, quantum physics.

Complimentarity The relationship between two complimentary quantities, such that, the more we know about one quantity, the less we know about the other. For example, the more we know about a particle's location, the less we know about its momentum.

Correlation This is in reference to the quantum eraser experiment, in which we mark the dots on the detection screen. We can then observe wave-like behavior when we look at the pattern of the dots. Different marks are used, allowing us to determine whether the atom left the top slit (traveling in a wave) or the bottom slit (also traveling in a wave). When we *correlate* the two different marks (one mark for particles we know left the top slit, and another mark for the bottom slit), we observe two separate sets of interference fringes.

Detectors, which-way Used in quantum erasure experiments. If we get a count from a detector in a cavity with the upper slit, we know a particle traveled through the upper slit. Likewise, if there

is no click, then we know the particle traveled through the bottom slit.

Energy, conservation of A principle stating that the total energy of an isolated system remains constant regardless of changes within the system.

Engine efficiency This book often refers to hypothetical reversible engines. In such engines, we are concerned only with the temperature difference between hot and cold reservoirs. In mechanical terms, the more heat energy we can put in to an engine (which we then have to cool), the more efficient the engine is. Thus, based on the type of fuel they use, one engine may be more powerful than another, but they all have the same efficiency.

Entanglement Entanglement refers to the correlation of particles in the stage before they hit the detection screen. For example, if we get a "click" from a photon in the top cavity, we know the particle went through the top slit; the particle and the photon are entangled.

Entropy The tendency of all systems to go toward the state of greatest disorder. This book frequently addresses information entropy; specifically, the definition of entropy as it applies to our ability to track the location of an atom in a hypothetical engine. As disorder increases, it becomes more difficult to track the atom.

EPR (Einstein–Podolsky–Rosen) paradox (or problem) A thought experiment which claimed that the result of a measurement performed on one part of a quantum system can have an instantaneous effect on the result of a measurement performed on another part, regardless of the distance separating the two parts. Einstein, Podolsky, and Rosen introduced the thought experiment in a 1935 paper to argue that quantum mechanics is not a complete physical theory.

Fixed bar magnet In quantum eraser experiments, we use fixed bar magnets. They allow us to observe both wave-like and non-wave-like behavior.

Floating bar magnet A floating bar magnet recoils when hit by an atom. This causes the atom to scatter, hitting the detection screen at a tangent because it bounced off the "floating" slit. Einstein and others have used floating bar magnets in experiments designed to disprove quantum mechanics. They were proved wrong, as information showed itself to be the answer to the peculiarities of quantum mechanics.

Free will The place of man to affect his own destiny. In quantum physics this free will is inexplicably related to the results of experiments involving atoms, photons, and electrons. Our conscious decisions affect reality in life – and in the lab.

Heat/heat source A form of energy associated with the motion of atoms or molecules and capable of being transmitted through solid and fluid media by conduction, through fluid media by convection, and through empty space by radiation.

Information, which-path When we have which-path information, interference fringes are wiped out. This means that the particles are traveling in straight lines.

Interference fringes Interference fringes are represented by the bright–dark areas on detection screens. They show that particles are traveling in wave-like patterns.

Interference (of waves) In quantum physics experiments, wave interference from light (photons) or atoms is seen in the patterns on detection screens.

Isothermal compression In an internal combustion engine, this is the exhaust stroke. In accordance with thermodynamic law, this is the process by which we restart the cycle. In information theory (e.g., Szilárd's single-atom engine), this is the process in which we expend energy by resetting the register.

Joule (J) The SI unit of electrical, mechanical, and thermal energy.

Kelvin (K) The SI unit of absolute temperature. Water freezes at 273.15 K and boils at 373.15 K.

Laser Optical application that produces an intense beam of light, composed of only one frequency.

Maser The forerunner to the laser. The maser uses excited atoms to convert pulses into microwave frequencies.

Maxwell's demon An imaginary being, used by science to describe subtle aspects of physics. Originally thought of by the English scientist James Maxwell, the demon was characterized as a fictional being who had the job of separating hot from cold molecules. He was developed to aid in science's lack of understanding of thermodynamic principles. He was "slain" by the advent of thermodynamics, only to rise again and again in the face of more subtle aspects such as those involved in quantum physics.

Micromaser A maser that operates with only one atom in it (the one-atom maser). It is a theoretical device helping us to understand the behavior of particles, just as Szilárd's single-atom engine helps us to understand information entropy.

Momentum A measure of the motion of a body, equal to the product of its mass m and velocity v (denoted p).

Negative entropy Often referred to as negentropy.

Negentropy The opposite of entropy. A state of extreme order. This is usually the case at the beginning of a process.

Observer In a quantum eraser experiment, the observer is a person observing which way the particle went. This conscious decision to observe erases wave-like behavior, which is evidenced by the lack of interference fringes. The lack of an observer will yield the opposite result.

Particle-like The opposite of wave-like behavior. Particle-like behavior is when particles travel in a straight line like machine gun bullets. The pattern they leave on a detection screen is one without interference fringes. The presence of an observer is necessary to cause this.

Particles Particles of light (photons), particles of matter (atoms), and electrically charged particles (electrons) are all used in quantum experiments.

Perpetual motion machine (of the second kind) A hypothetical machine that extracts useful work from a single thermal bath (e.g., the ocean).

Photoelectric effect Ejection of electrons from a metal by incident light.

Photon The quantum of light, regarded as a discrete particle having zero mass, and no electric charge.

Quantum eraser An experiment in quantum physics in which we erase information we didn't actually have but *could have* had. Upon understanding the quantum eraser, the reader will also resolve the EPR "paradox." See Section 10.5.

Quantum physics Physical theory describing atoms and light. The deterministic laws of classical physics are replaced by a wave-like equation that predicts only probabilities of fundamentally random events. The name "quantum" arises from the fact that certain continuous quantities in classical physics (such as energy) become discrete, or quantized.

Recoiling slit A recoiling slit moves when struck by a particle. The particle recoils and the direction of recoil of the slit tells us which side of the slit was struck. Einstein and others used this experiment to argue against the understanding we have today about wave–particle duality. They claimed the recoil–ricochet effect was responsible for the loss of interference fringes rather than an observer.

SI The International System of units (involving the meter, kilogram and second).

Single-atom engine This is a hypothetical engine which helps us to understand the key to improved engine design. Studying the nature of an individual atom at work helps us to see clearly the natural laws of physics.

Steam engine The steam engine ran at about six percent efficiency before the laws of thermodynamics were employed in its design. Carnot helped to change this, allowing the design of steam engines to achieve much greater efficiency.

Stern–Gerlach apparatus The celebrated experiment of Otto Stern and Walther Gerlach involves sending a beam of spin $\frac{1}{2}$ particles through an inhomogeneous magnetic field. Each spin $\frac{1}{2}$ particle has a spin magnetic moment. The inhomogeneous magnetic field causes the particle to be deflected either up or down, and thus measures the magnetic moment of the particle.

Szilárd's engine A hypothetical single-atom engine, designed to illustrate the fundamental physics of entropy, work, and thermodynamics.

Thermodynamics The study of energy in the form of heat. The study of thermodynamics allowed for the advent of modern physics and quantum physics.

Thermodynamics, first law of This law states that heat added to work on a system equals the change in its internal energy. Hence, the greater the amount of heat, the greater the energy output. In engineering designs this means a good cooling system was essential, as it allows for a greater difference between hot and cold in the engine's operating temperatures.

Thermodynamics, second law of This law states that heat flows from hot to cold. It exemplifies the need for an exhaust (cooling) stroke on all heat engines. It makes impossible the existence of the so-called perpetual motion machine. It was once assumed that Maxwell's demon did this sorting of hot and cold molecules.

Transistor A small electronic device containing a semiconductor and having at least three electrical contacts, used in a circuit as an amplifier, detector, or switch.

Triangular and trapezoidal magnets When one magnet is triangular shaped and the other is trapezoidal, they exert non-uniform magnetic fields. This allows for the particles passing between them in an experiment to be randomly deflected.

Tripos Any of the examinations for the B.A. degree with honors at Cambridge University in England.

Two slits Two horizontal slits on an opaque screen that are used in quantum physics experiments. When we know that the particle went through the top slit or the bottom slit, we have which-way information. This wipes out interference fringes, indicating particle behavior. When we don't have which-way information, we have interference fringes, indicating wave-like behavior.

Uncertainty principle (or relation) Heisenberg was responsible for this. It is described in Fig. 8.4. It states that, the more we know of a particle's location, the less we know of its velocity. The scattering of particles off the narrow slits is the reason for the uncertainty. The narrower the slit, the greater the ricochet.

Wave function A mathematical function used in quantum mechanics to describe the probability of a particle being found at any given location.

Wave-like The tendency of light, particles of matter, and electrons to travel in wave-like patterns. An observer in a controlled experiment will inexplicably "erase" wave-like behavior.

Wave mechanics The quantum study of wave behavior, invented by Schrödinger. When atoms, photons or electrons travel in waves, they can bend around corners.

Wave-particle duality The ability of quantum particles to travel as waves or in straight lines as particles. The exhibition of both wave-like and particle-like properties by a single entity, such as both diffraction and linear propagation of light.

Which path Which way.

Wigner's friend Eugene Wigner developed the idea of a conscious observer who relayed information to us about which way (which path) the particle took, thus rubbing out interference fringes. See Section 9.1.

Citations, Comments and Further Readings

Chapter 1

Citations and Comments

[1] Most of the Pythagorean material cited comes from: K. Guthrie, *The Pythagorean Sourcebook and Library*, Phanes Press, 1987. See also Harding Samuelson, *Numerology*, Sherbourne Press, 1970.

[2] Arnold Herman, *To Think Like God*, Parmenides Publishing, Las Vegas, 2004.

[3] Michael Beer, *How Do Mathematics and Music Relate to Each Other?*, East Coast College of English, Brisbane, Australia, 1998.

[4] Material on π derived from: Peter Beckman, *A History of π*, St. Martin's Press, New York, 1971. I found the little ditty by Auden in the excellent book: Paul Nahim, *An Imaginary Tale: The Story of* $\sqrt{-1}$, Princeton University Press, Princeton, NJ, 1998.

[5] Brian Butterworth, *What Counts, How Every Brain is Hardwired for Math*, Free Press, 1999.

[6] 1996 Inter Nationes, Bonn–Bad Godesberg.

[7] Fred L. Wilson, Rochester Institute of Technology, www.rit.edu/flwstv/plato.html.

Further Reading

Malcolm E. Lines, *On the Shoulders of Giants*, Institute of Physics Publishing, Bristol, UK, and Philadelphia, 1994. Lines follows the history of science with an eye to the ways mathematics has nurtured physical science from the Greeks to the geeks.

Chapter 2

Citations and Comments

[8] Owen Gingerich, "God's goof", in *Science and Religion: Are They Compatible?*, Ed. Paul Kurtz, Prometheus Books, Amherst, NY, 2003.

[9] Joshua Gilder, and Anne-Lee Gilder, *Heavenly Intrigue*, Doubleday, 2004, p. 36.

[10] Ibid., p. 70.

[11] Owen Gingerich, Private communication.

The Demon and the Quantum, Second Edition. Robert J. Scully and Marlan O. Scully

ISBN 978-3-527-40983-9

[12] Ibid.
[13] Joshua Gilder, and Anne-Lee Gilder, *Heavenly Intrigue*, Doubleday, 2004, p. 120.
[14] Ibid., p. 202.
[15] Ibid., p. 200.
[16] Owen Gingerich, Private communication.
[17] Paul Kurtz (Ed.), *Science and Religion: Are They Compatible?*, Prometheus Books, Amherst, NY, 2003.
[18] E. N. da G. Andrade, *Sir Isaac Newton, His Life and Work*, Doubleday Anchor, 1954, p. 3.
[19] Michael White, *Isaac Newton. The Last Sorcerer*, Helix Books, Reading, MA, 1990.
[20] Ibid., p. 4.
[21] Ibid., pp. 327–342.
[22] John Henry, *Knowledge is Power*, Totem Books, 2003.
[23] *New Organon*, II, Aphorism 20.
[24] Marlan O. Scully, Private communication.
[25] David Lindley, *Degrees Kelvin*, John Henry Press, Washington DC, 2004, gives a nice discussion of heat and its measurement.

Chapter 3

Citations and Comments

[26] Hans Christian Von Baeyer, *Warmth Disperses and Time Passes*, Random House, New York, 1999, p. 98.
[27] John Gribbin, *The Scientist*, Random House, 2004.

Further Reading

S. R. Carnot, *Reflections on the Motive Power of Heat*, translated by R. Thurston, American Society of Mechanical Engineers, New York, 1943. This excellent little book should be read by anyone who has an interest in the development of science.

John B. Fenn, *Engines, Energy, and Entropy: A Thermodynamics Primer*, Global View Publishing, Pittsburgh, 2003. Nobel laureate Fenn develops thermodynamics in his own unique witty way. The book is good fun and good science.

Another excellent popular treatment of heat engines and Carnot's contributions to thermodynamics is to be found in: David Goodstein, *Out of Gas*, W. W. Norton, 2005.

A standard text on introductory thermodynamics is: Francis Sears, *Thermodynamics*, Addison Wesley.

For an outstanding, more advanced treatment see: I. Puri and K. Annamalai, Advanced Thermodynamics Engineering (CRC, 2001).

Chapter 4

Citations and Comments

[28] Hans Christian Von Baeyer, *Warmth Disperses and Time Passes*, Random House, New York, 1999.

[29] Footnote in Richard P. Feynman, *The Feynman Lectures on Physics*, Harperaudio Publishers, 2006.
[30] Bernard Pullman, *The Atom in the History of Human Thought*, Oxford University Press, New York, 1990.
[31] Leon Brillouin, *Science and Information Theory*, 1962.
[32] I thank Goong Chen, Professor of Mathematics, Texas A&M University, for this observation.
[33] David Lindley, *Boltzmann's Atom*, The Free Press, 2001, p. 224.
[34] B. De Witt, "God's rays," *Physics Today*, 9 September 2005.
[35] C. S. Lewis, *A Terrible Mercy* or *Surprised by Joy*.
[36] E. Schrödinger, *What Is Life, With Mind and Matter and Autobiographical Sketches*, Cambridge University Press, Cambridge, UK, 1944. Also see: E. Schrödinger, *My View of the World*, Ox Bow Press, Woodbridge, CT, 1951.
[37] H. Margenau, *The Miracle of Existence*, Ox Bow Press, Woodbridge, CT, 1984.

Chapter 5

Citations and Comments

[38] Harvey S. Leff, and Andrew F. Rex (Eds.), *Maxwell's Demon: Entropy, Information, Computing*, Princeton University Press, Princeton, NJ, 1990.
[39] M. von Smoluchowski, Lecture notes, Leipzig, 1914 [as quoted by L. Szilárd, *Z. Phys.* **53**, 840–856 (1929)].
[40] R. Landauer, *IBM Journal of Research and Development*, **44**(1/2), 261 (2000).
[41] C. H. Bennett, "Demons, engines and the second law," *Scientific American*, November 1987, p. 88.

Chapter 6

Citations and Comments

[42] Harvey S. Leff, and Andrew F. Rex (Eds.), *Maxwell's Demon 2: Entropy, Classical and Quantum Information, Computing*, Institute of Physics Publishing, Bristol, UK, 2003.
[43] Ibid.

Further Reading

R. P. Feynman, R. B. Leighton, and M. Sands, *The Feynman Lectures on Physics*, Addison-Wesley, 1965.
Hans Christian Von Baeyer, *Taming the Atom: The Emergence of the Visible Microworld*, Random House, New York, 1992.
Jennifer Ouellette, with foreword by Alan Chodos, *Black Bodies and Quantum Cats: Tales from the Annals of Physics*, Penguin Books, London, 2005.
P. C. W. Davies, and J. R. Brown, *The Ghost in the Atom: A discussion of the Mysteries of Quantum Physics*, Cambridge University Press, UK, 1986.
Sam Treiman, *The Odd Quantum*, Princeton University Press, Princeton, NJ, 1999.

Kenneth W. Ford, *The Quantum World: Quantum Physics for Everyone*, Harvard University Press, Cambridge, MA, 2004.

Chapter 7

Citations and Comments

[44] Dirk ter Haar: *Elements of Thermostatistics*, Holt, Reinhart, 1966, p. 224.

Chapter 8

Citations and Comments

[45] Walter Moore, *Schrödinger Life and Thought*, Cambridge University Press, UK, 1989, p. 242.
[46] Henry A. Boorse, *The World of the Atom*, Basic Books, New York, 1966.
[47] T. Young, *A Course of Lectures on Natural Philosophy and the Mechanical Arts*, Vol. I, Printed for J. Johnson, London, 1807, Fig. 267.
[48] Walter Moore, *Schrödinger Life and Thought*, Cambridge University Press, UK, 1989, p. 272.
[49] David C. Cassidy, *Uncertainty: The Life and Science of Werner Heisenberg*, W. H. Freeman, 1992, p. 16.
[50] Ibid., p. 85.
[51] Ibid., p. 347.
[52] Ibid., p. 341.
[53] John Cornwell, *Hitler's Scientists*, Viking, 2003.
[54] Brian Greene, *The Fabric of the Cosmos*, Knopf, 2004.
[55] Ibid.

Further Reading

G. Greenstein and A. Zajone, *The Quantum Challenge* (Jones and Bartlett, Mass: 1977). See also: A. Tonomura, et al., *Am. J. Phys.* **57**, 117 (1989) and A. Hobsson, *Am. J. Phys.* **73**, 630 (2005).

Chapter 9

Citations and Comments

[56] Eugene Wigner, in *The Scientist Speculates*, Ed. I. Good, Basic Books, 1962.
[57] Brian Greene, *The Fabric of the Cosmos*, Knopf, 2004.

Further Reading

Jim Baggott, *Beyond Measure: Modern Physics, Philosophy, and the Meaning of Quantum Theory*, Oxford University Press, UK, 2004.

Chapter 10

Citations and Comments

[58] John S. Bell, "Against 'measurement'," *Physics World*, August 1990, pp. 33–35.
[59] Henry F. Schaefer, III, *Science and Christianity: Conflict or Coherence*, The Apollo Trust, 2003.
[60] R. P. Feynman, R. B. Leighton, and M. Sands, *The Feynman Lectures on Physics*, Addison-Wesley, 1965.
[61] Paul Weiss, "Wave or particle? Heisenberg, take a hike!," *Science News*, 5 September 1998.
[62] Ibid., p. 149.
[63] Ibid.
[64] E. H. Walker, *The Physics of Consciousness*, Perseus Publishing, Cambridge, MA, 2000.
[65] Jim Baggott, *Beyond Measure: Modern Physics, Philosophy, and the Meaning of Quantum Theory*, Oxford University Press, UK, 2004.
[66] A. A. Harkavy, *Human Will – The Search For Its Physical Basis*, Peter Long, 1995.
[67] A. Compton, *Science*, **74**, 172 (1931).
[68] Gregg Herken, *Brotherhood of the Bomb*, Henry Holt, New York, 2002.
[69] Judith Shodery, *Memoirs, Edward Teller*, Perseus Publishing, Cambridge, MA, 2001.
[70] Paul Davies, *The Mind of God*, Touchstone, 1993.
[71] Quoted in: C. B. Boyer, *A History of Mathematics*, Wiley, 1968.
[72] P. Zoeller-Greer, "Divine anthropic principle," *Perspectives on Science and Christian Faith*, 52, March 2000, p. 8.
[73] M. O. Scully, and K. Drühl, "Quantum eraser: a proposed photon correlation experiment concerning observation and 'delayed choice' in quantum mechanics," *Phys. Rev. A*, **25**, 2208 (1982).
[74] G. Schroeder, *The Science of God*, The Free Press, New York, 1997.
[75] F. Hoyle, in *California Institute of Technology Alumni Publ., Engineering and Science*, November 1981, p. 12.
[76] Paul Davies, *The Mind of God*, Touchstone, 1993.
[77] Ibid., p. 162.
[78] H. Weyl, *The Open World*, Ox Bow Press, Woodbridge, CT, 1931.
[79] T. Rothman, and E. C. G. Sudarshan, *Doubt and Certainty*, Perseus Books, Reading, MA, 1998.
[80] W. Heisenberg, *Across the Frontiers*, Ox Bow Press, Woodbridge, CT, 1990.
[81] Owen Gingerich, "God's goof", in *Science and Religion: Are They Compatible?*, Ed. Paul Kurtz, Prometheus Books, Amherst, NY, 2003.
[82] Louise Ropes Loomis, *Plato, Five Great Dialogues*, Granercy Books, New York, 1969.
[83] Ibid.

Further Reading

Siegfried, Tom, *Bit and the Pendulum: From Quantum Computing to Quantum Theory – The New Physics of Information*, John Wiley and Sons, Inc., 2000.

Fred Alan Wolf, *Mind Into Matter: A New Alchemy of Science and Spirit*, Moment Point Press, Needham, MA, 2001.

Roger Penrose, *The Emperor's New Mind*, Oxford University Press, UK, 1999.

John Polkinghorne, *Belief in God in an Age of Science*, Yale Nota Bene, Yale University Press, New Haven, CT, 1998.

Allan A. Harkavy, *Human Will: The Search For Its Physical Basis*, Peter Lang, New York, 1995.

Hugh Ross, *The Creator and the Cosmos: How the Greatest Scientific Discoveries of the Century Reveal God*, Navpress, Colorado Springs, CO, 1995.

Charles B. Thaxton, Walter L. Bradley, and Roger L. Olsen, *The Mystery of Life's Origin: Reassessing Current Theories*, Lewis and Stanley, Dallas, TX, 1984.

Hugh Ross, *Creation and Time: A Biblical and Scientific Perspective on the Creation-Date Controversy*, Navpress, Colorado Springs, CO, 1994.

John Polkinghorne, *The Way the World Is*, William B. Eerdmans, Grand Rapids, MI, 1983.

Index

The Demon and the Quantum, Second Edition. Robert J. Scully and Marlan O. Scully

ISBN 978-3-527-40983-9

f

g

h